KING OF SEA DIAMONDS

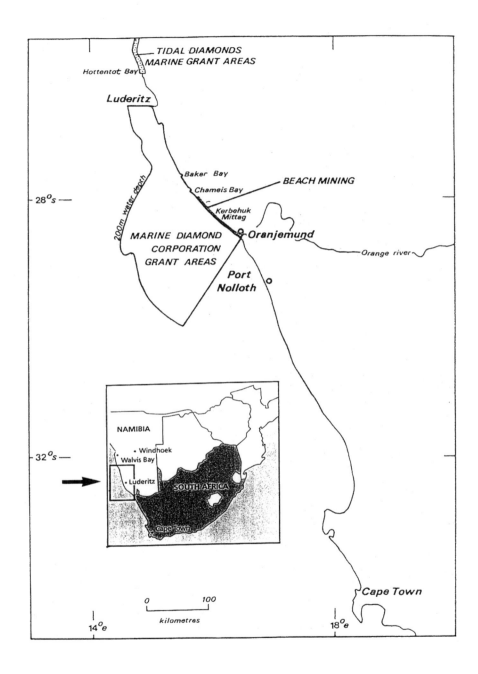

The location of the offshore concession acquired in the '60s by Collins's Marine Diamond Corporation. The overall marine diamond areas of Namibia are shown on the inset map.

KING OF SEA DIAMONDS

The saga of Sam Collins

Roger Williams

W.J. FLESCH & PARTNERS

October 1996
ISBN 0 949989 69 X

© Roger Williams 1996

W.J. FLESCH & PARTNERS (PTY) LTD
P.O. Box 3473 Cape Town
Telephone 27 21 461-7472
Fax 27 21 461-3758
E-mail sflesch@aztec.co.za

Typesetting and repro by Photoprints (Pty) Ltd
and printed and bound in the
Republic of South Africa
National Book Printers
Drukkery Street, Goodwood, Western Cape

CONTENTS

Foreword 7
Acknowledgements 9
Introduction 11
1. Into oil and pipelines, in Texas 13
2. Keeble tells Collins about sea diamonds 17
3. Sammy buys an offshore concession 29
4. *Emerson K* – and mutiny at Luderitz 42
5. *Barge 77* leads the way; astounds critics 51
6. Little-known facets of a rough diamond 63
7. 'A great leader – and a great boss' 75
8. On the rocks – on a desert coast! 87
9. Back in the hunt – with bigger barges 93
10. Aircraft acquired; more setbacks afloat 101
11. Life in the floating mining camps 116
12. Vema Seamount, Tretchikoff – and Verwoerd 125
13. The great sea diamond rush gains momentum 132
14. De Beers takes over 138
15. Years later, 'the Collins lamp is relit' 150
16. The Proof of the Plum Pudding 157
17. Sammy, alas, 'only scratched the surface' 167
Addendum 171
Index 172

Sam Collins with some of his sparkling "ice" – a parcel of 10 000 carats of gem diamonds sucked from the sea-bottom.

Lt-Cdr Peter Keeble, OBE, former naval diver, author and businessman. He told Collins about diamonds in the sea.

FOREWORD

Ivan R M Prinsep
Chairman,
Ocean Diamond Mining Group
Montreux, Switzerland
June 1996

ONLY a few years ago, anyone at all would have said that the only diamonds existing in the sea would be those that, in the light of the moon or of the sun, sparkle so beautifully on the ocean's surface.

It took a courageous, inventive and hard-working Texan, Sammy Collins, to prove that a large quantity of fine gemstones could be garnered from one particular part of the world's sea-floor – off the south-west coast of Southern Africa.

Many people involved in the diamond industry owe him a debt of gratitude. This book is a tribute to him, and to his facing and overcoming the many challenges and obstacles that lay in his path.

Like many a true pioneer, Sammy Collins did not live to see the eventual rapid growth of the industry he started; a growth which now increasingly captures the imagination of all those studying new technology, the oceans and particularly mining methods that are not harmful to the ecology of our world.

The writer of this book, Roger Williams – himself an ex-naval man and familiar with the sea – has said that in many years of writing and interviewing he has seldom come across one person about whom so many people have had not only stirring stories to tell but at the same time have also evinced admiration, respect and fondness.

Sammy was indeed admired and respected by many – also loved by some – and the mere mention of his name has brought warm smiles of remembrance to those who have spoken of him.

To be recalled in this way by so many who have contributed, through Roger Williams, their recollections for the compilation of this book is perhaps the greatest testimonial to any man.

It is hoped that the reader will share in the excitement of this unusual story, involving one of the greatest hoards of gemstone diamonds in the world and the challenge of mining them from the sea.

To ladies reading this book, one might say that if they already treasure diamonds in their possession, they may well prize them all the more after learning of the tremendous battles involved in getting sea diamonds from the floor of the stormy, open Atlantic Ocean to their final places of rest, on rings and necklaces.

Captain George Foulis, first master of the Emerson K. *He later became Sam Collins's overall marine superintendent.*

Emerson K, *formerly the Admiralty tug* HMS Marauder, *played a pioneering role in the mining of sea diamonds.*

ACKNOWLEDGEMENTS

THE author wishes to express his sincere appreciation and gratitude to all those who have contributed to the compilation of this biography, by way of personal reminiscence and/or photographs.

Without their help it would not have been possible to produce a book of this kind, and while it is well-nigh impossible to acknowledge each source of information individually, the names mentioned hereunder are those of persons whose contributions have been particularly helpful.

Special mention must be made of my research assistant Mr Neville Prehn, who spent several months delving into archival material at the South African Library, Cape Town, and at Newspaper House and the Cape Town offices of Times Media Limited and other newspaper and magazine offices. He was also able to establish valuable contacts in the late Sam Collins's home state, Texas, in the USA, and to obtain important background information from civic libraries at Beaumont and Port Lavaca, both in Texas.

Others to whom I would like to convey my warmest thanks for their help and co-operation are:

Mr Tom Kilgour, Mr Francois Hoffman, Mr Bill Webb, Mr Hans Abel, Mr Gary Haselau (many of whose photographs including the front-cover picture have been used in this book), Dr Piet Neethling, Captain George Foulis, Mr Piet Beukes, Messrs Joe and Andrew Horne, Ms Marilyn Martin, Captain "Okkie" Grapow, Mr Hans Duyzers, Mr Danny Ipp, Mr Tony Brewer (Editor of the *Mining Journal*, London), Mr Tony Heard (former Editor of the *Cape Times* and at one time Cape Editor of the *Financial Mail*), Mr Vladimir Tretchikoff, Mr Eddie Redgrave, Mr Jock Carr, Mrs Pam Greenshields, Mr Socrates Vartsos, Captain Hugh Pharoah of Singapore Airlines, Mr John Bevil Rudd, Mr Steve Phelps, Mr Geoff Grylls, Mr George Brown, Mrs Babs Vivier, the late Mr Sidney Kagan and Mr Johnny de Olim.

The 50-ft fishing boat Karibib, *which Johann Vivier used to extract the first diamonds found in the deep sea off SWA.*

Johann Vivier, the Beaufort West farmer who sold his mining concession on the SWA coast to Sam Collins, in 1961.

Punching through heavy swells, a small work-boat heads seaward during exploratory work off the hostile SWA coast.

INTRODUCTION

"There is no other commodity with such an irresistible fascination as diamonds. People will risk all for them. They will betray, cheat, lie, deceive and murder. They will forfeit honour, friendship and loyalty. They will suffer privations and danger. For diamonds make dreams a reality. They bring prestige and power. They open doors to presidents and kings. They magically transfer poor men into millionaires."

— Olga Levinson, in her book 'Diamonds in the Desert',
the human and poignant story of August Stauch.

FOR four hectic and exciting years in the 1960s a stocky Texan oilman, Samuel Vernon Collins, played a pioneering role in the mining of diamonds from the sea off the desert coast of Namibia, then known as South West Africa.

A hard-driving, hard-living and innovative man, Sammy Collins wrote a whole new chapter into the romantic and ever-fascinating story of the search for and the recovery of gemstones.

He succeeded where others had flinched or failed, and like Barney Barnato and Hans Merensky he became a legend in the mining industry. A close associate has likened working for Collins to "living next to a volcano". The Texan has also been referred to as "a great leader, and a great boss".

Rough, tough and ever-demanding of those who worked for him, he gained their respect and loyalty through fair-mindedness, generosity and close personal attention to their needs.

Taking up the challenges thrown at him by the salty elements off a desolate coast, he confounded the critics and the cynics by proving the presence of large concentrations of diamonds on the seabed – and by succeeding in mining them in payable quantities.

Not only that; he also injected new life into the flagging economy of the area from which he operated, Cape Town and the Western Cape. His Marine Diamond Corporation generated new industrial and business activity, created new job opportunities and generally "woke up a sleepy Cape Town".

Wherever the restless and dynamic Sammy Collins went, he left his mark, not only at the Cape but also in other parts of Southern Africa and of the world. And his flamboyant lifestyle left a wealth of colourful stories and anecdotes.

This book is an attempt to recapture some of the colour and flair of the brief but bountiful reign of the man who came to be known as the "king of sea diamonds".

Metallurgist Bill Webb (left) and geologist Francois Hoffman look for diamonds in gravel sucked up by Emerson K.

CHAPTER ONE

Into oil and pipelines

A Texan tycoon in embryo

ONE needs to look no farther than his birthplace to understand why Samuel Vernon Collins became the human fireball that he was.

For it was here, at Beaumont in southern Texas, that the first great oil gusher in North America – the Lucas gusher of the Spindletop field – blew in with a roar in 1901, pouring fortunes into muddy streets and launching a great oil industry.

Sammy was born into this dynamic world of oil and oilmen on July 23, 1913. His parents were Frank and Ellie Collins.

Frank worked in an oil refinery, so it was hardly surprising when young Sammy showed an early interest in the exciting opportunities offered in the production of a substance for which there was a fast-growing global demand.

Beaumont, a relatively small place on the River Neches east of Houston, had a population of a mere 30 000 when young Sammy went to school there. Yet this once lonely log-cabin settlement, a place of gnarled oaks and wild magnolias, and redolent of the Deep South, came to be known as the "city of oil" in the oil-mad boom at the turn of the century.

Sammy spent the first 15 years of his life in Beaumont, and while there he developed a love of water sports, and of the sea. "We practically lived in the Neches River," he recalled in later life. "I was a champion swimmer, and a lifeguard, and I also became a swimming and diving instructor."

The Collins family moved to Corpus Christi in West Texas about 1928, when Frank got a job as a pipeline station operator there. And during the Great Depression of the early '30s, Sammy sold magazines door-to-door, to augment the paltry family income and to earn pocket-money for himself.

He first started working on pipelines when he was 17 and, like a true Texan, he became more and more involved in the oil business – and he did well at whatever job he was given.

Stockily built and physically strong, he impressed his superiors as a hard-driving, dedicated worker who was also innovative, with leadership and entrepreneurial qualities.

He worked as a lowly "roustabout" on oil rigs for a while, and he later defined this term as "a guy who runs around with a monkey-wrench, getting machines to go and keeping 'em thataway".

Then he worked as a "roughneck", or oil-driller, but it was not long before his skills brought their own reward, and Sammy rose to become a "farm boss" at Corpus Christi. This is the term used for a foreman over a producing oil-well. He did that for a couple of years, then became a "tool-pusher", or foreman over a drilling rig.

As he went he learnt, and as he learnt he began evolving techniques of his own. An inventive man, he patented many of his ideas – and began reaping financial rewards from processes that were adopted for regular usage in oil mining.

By the time World War II broke out, in September 1939, the stripling from Beaumont had, at the age of 26, become a drilling superintendent. This meant he had control over all the tool-pushers at a number of drilling rigs. He was now making about 800 dollars a month, with all expenses and a car thrown in.

Young Sammy Collins had "arrived" in the oil business.

"I remained a drilling superintendent all through the war, as this was considered a vital defence job," he later told a prominent Cape Town journalist and editor, Piet Beukes, in the presence of Emerson Kailey, Sam's close friend and London representative.

"We'd probably have lost the war!..."

At this point in the interview Kailey interposed, with a chuckle: "If Sammy had been drafted, we'd probably have lost the war!"

At the end of 1944, Sam Collins formed his own oil-field contracting company, which did the service work for the oil-field drilling and production companies. "I went out and gathered a lot of Mexican 'naturals', about 400 of them, to keep the oil wells producing. This was a service for all the oil companies, as there was a great scarcity of labour.

"It was at that time, in 1945, that one of my clients had drilled offshore wells in Matagorda Bay in Texas, and they had no-one to do the service work out there. They got me to lay the pipelines to the well in the sea.

"It was the first time that an oil pipeline had been laid in the sea. We found that no-one knew anything about working in the sea. We also knew nothing. It was then that all the big companies were moving into the sea to start wells.

"We formed a new company to service the offshore areas. Eventually, I formed three companies and they were very successful, earning a lot of money. The first of these, Collins Construction Company (based at Port Lavaca, Texas) became the parent company of the whole outfit.

"In 1951, when the natural gas industry began to blossom, it was decided to take the Texas gas to New York and Chicago. The biggest headache for the gas companies was getting their pipelines across the big rivers. They came to us, to do the river crossings.

"By this time we had developed and patented the Collins submarine pipeline trencher, which works on the principle of cutting trenches with powerful water jets along sea and river beds, then burying the pipelines in them.

"My partners in the three companies did not want to expand, and we agreed to dissolve the companies – and I went on alone."

It was recorded in Texas at the time that Collins completed a petroleum pipeline operation across the Mackinac Straits in Michigan, using his jet-trencher, special winches and pipeline launching systems "which later were used in solving problems in the way of completing Transcontinental's Narrows, and the Uptown and Peekskill crossings of the Hudson River".

It was further recorded that, in 1956, Collins's pipeline firm set a new world record by laying 60 000 feet of 10,75-inch concrete-coated pipe across Corpus Christi Bay in 80 hours.

Collins Submarine Pipelines had by then started receiving calls from overseas, and the firm's first job out of America was in India, where service pipelines were laid for the Bombay Port Trust, in the Bay of Bombay, using loading trenchers.

The next job was in Ghana, for an oil pipeline in the ocean and, said Sammy, "after that we got pipeline contracts in all parts of the world, worth a total of about £35 million."

Collins started his submarine pipeline business at Port Lavaca with a loan of $2500 from a bank in Texas, and within a few years he was able to boast of owning about £5 million worth of equipment in scattered parts of the world.

A world authority on submarine pipelines

Apart from becoming a world authority on the construction and laying of submarine pipelines, he also owned companies producing oil in Texas and Louisiana – and had amassed a personal fortune.

At Port Lavaca, a natural port on Calhoun County's Matagorda Bay, Sammy found the ideal base for his pipeline and associated activities. The coastal county is located near the centre of the 375-mile Texas Gulf Coast, and is the home of a growing deepwater port and more than 100 miles of inland waterways.

Only 100 miles south-west of Houston, Calhoun County had

become an important centre for oil and gas exploration in Collins's time there. And, situated about midway between the Atlantic seaboard and the coast of California, it is also a mere 200 miles from Mexico, across the Rio Grande.

Although 18 million people live within a 300-mile radius of it, Calhoun County itself had a total population of a mere 20 000 as recently as 1992 – and of this number, little more than 11 000 lived at Port Lavaca.

The tourism authority in Calhoun County uses as a slogan: "We're the best-kept secret on the Texas Coast!"

By 1957, the Collins Submarine Pipeline Company had perfected a technique of operating under the sea in all conditions and had learnt how to mine gravel by using airlifts.

"See the scars on the back of my hands," Sammy said to his interviewer. "The jets are very powerful, and I often fell down while working with the pipes and with the jets under the oceans, and they cut the skin right off my hands."

Collins had himself become a qualified and highly-experienced diver by this time. He had worked in water-depths up to 225 feet, and had suffered "the bends" twice.

He was trapped on a number of occasions while working under the sea, and he had a particularly close call when there was a cave-in under the Hudson River in New York.

He recalled later: "We were working in 100 feet of water and had dug a ditch about 18 ft deep. I was working on a machine, and the walls of the ditch fell in on top of me. I was trapped.

"I was in full-dress diving suit and had my air-pipe. They attached a water hose onto my lifeline and I managed to pull the lifeline and the water hose through the mud towards me and finally, after about four hours, I worked myself out of that mess."

Collins said in his interview with Piet Beukes that it must be understood that the only talent he had was for mechanics. "It is not for business. All the money we ever made was made out of new methods, and evolving new techniques for underwater work. We had to find a way to do things that no-one else could do.

"I was equipped to do this because I was a good swimmer, a professional diver and, on the engineering side, I had taken a course in engineering. It was a two-year course which I finished in six weeks."

It was perhaps inevitable that one former professional deep-sea diver of note should meet another at some point in their careers.

Particularly when they were both also noted for their personal dynamism, for their unbounded energy and daring – and for their genius as inventive engineers.

Enter Lt-Commander LAJ "Peter" Keeble, OBE a South African hero of naval salvage in the Mediterranean during World War II.

CHAPTER TWO

Keeble gets Collins interested in sea diamonds

*"Full many a gem of purest ray serene
The dark unfathom'd caves of ocean bear..."*

— Gray's Elegy

JAMES DUGAN, a best-selling author and a friend of Sam Collins, was instrumental in getting Peter Keeble on Collins's payroll in the late 1950s – and this led to a dramatic change of course in the Texan's career.

Dugan had recently produced his book *The Great Iron Ship*, the astonishing saga of Brunel's *Great Eastern*, the massive and revolutionary vessel that laid the Atlantic cable in the 19th century. And in 1957 Keeble, who had become a wartime OBE for his part in clearing blocked harbours in the Mediterranean and for entering a sunken U-boat to recover a top-secret device, had had his memoirs, *Ordeal by Water*, published in London.

Collins later recounted how Keeble came into the pipeline picture – and how the South African got the Texan interested in sea diamonds.

"I was in Texas and Jim Dugan, the famous writer, phoned me from New York and said he had this guy with him from South Africa who had some patented work for laying undersea pipelines in South Africa – but he had no money.

'I said Keeble should see Emerson Kailey at our London office. This happened, and Kailey put him on the payroll.

"Kailey then set about creating a new company, Collins Submarine Pipelines Africa (Pty) Limited, and through Peter Keeble he also got a saltpan concession.

"Then, one evening at our office in London – 18 Park Lane – Kailey, Keeble and I had drinks together and Keeble, who is a good story-teller, mentioned there was a good possibility of diamonds being found off the coast of South West Africa.

"I said I had read about this in a magazine – it could have been the *Reader's Digest*. I was in Texas at the time, and that article damn sure did make an impression on me. The reason I didn't follow it up is that we had not yet developed a technique that would allow us to work in heavy seas.

"Peter told us a hundred stories about people taking out diamonds from the sea, and we saw a confidential report to the directors of De Beers, signed by Sir Ernest Oppenheimer. This said diamonds were most certainly in the sea, but there was no way of getting them out.

"Keeble also quoted Oppenheimer's associate Alpheus F Williams as having written that if the sea were to dry up, diamonds would be found in many places on the ocean floor.

"Kailey said the whole thing is crazy, and let's talk business about pipelines instead."

Soon after this meeting in London, Sam Collins went to the Persian Gulf to supervise the installation by his company of the longest submarine pipeline ever built. This was 38,6 km (24 miles) long and 76 cm in diameter, and it reached from the Iranian mainland to an offshore island.

Back in London Collins was contacted by Abe Bloomberg MP, a leading Cape Town attorney, parliamentarian and entrepreneur. Bloomberg and others, including another well-known parliamentarian, Captain GHF "Kappie" Strydom, had interests in certain diamond concessions in the vicinity of the Orange River mouth.

Having made no headway with De Beers, which initially was not convinced that diamonds could be recovered from the sea in payable quantities, Bloomberg approached Collins on Keeble's advice – and told him there were positive indications of gem diamonds aplenty on the seabed off SWA.

Collins later also met Colonel Jack Scott, chief executive of the General Mining and Finance Corporation, who shared the Texan's

Collins, flanked by his security chief Rex Redelinghuys (left) and Bill Webb, scans some of the earlier sea diamonds.

enthusiasm and his interest in mining sea diamonds.

Collins needed no further persuading. Always amenable to new ideas and new challenges, he set his sights on finding ways of recovering gem diamonds from the sea in payable quantities – and he went for it, with Texan resolve and tenacity.

The story of diamonds thus far

The story of diamonds in South Africa goes back to 1866. The first one was found on the farm De Kalk in the Hopetown district on the Orange River, near Kimberley.

The diamond was picked up by Erasmus Stephanus Jacobs when he was a boy of 15. It became part of the collection of beautiful pebbles that the Jacobs children had assembled for their games of "klippie".

Several people at Hopetown and Colesberg took an interest in the beautiful and quite remarkable "pebble" found by young Jacobs, and after the acting civil commissioner at Colesberg, Lorenzo Boyes, discovered he could use the stone to scratch the initials of a local store – Draper & Plewman – on a window-pane of the shop, he sent it to Grahamstown for analysis.

Dr Guybon Atherstone, the mineralogist at Grahamstown, pronounced after a thorough examination that this was "a veritable diamond", weighing 21,5 carats and worth £500. After this had been confirmed in London, the diamond was acquired, for £500, by the Governor of the Cape at the time, Sir Philip Wodehouse.

Later diamond discoveries at Kimberley, Colesberg and other parts of South Africa became the catalyst for industrial growth in the country, earning huge sums in foreign exchange and providing employment for thousands.

Until overshadowed by gold, the mining and marketing of diamonds helped to lift the country out of a slough of agricultural poverty, and attracted dynamic entrepreneurs from all over the world.

The Big Hole

In 1880, as the diggings at Kimberley reached ever-greater depths in what came to be known as the Big Hole, and with strong indications that diamond-mining in South Africa had a long and viable life ahead of it, the great empire-builder Cecil John Rhodes and his associate Charles Rudd formed the original De Beers Mining Company after buying up a number of small claims.

The company was named in honour of the family headed by the Voortrekker Johannes Nicolaas De Beer, owner of the farm Vooruitzicht on which Kimberley now stands.

This, and other consolidating developments at the time, had the effect of putting the diamond market on a more secure footing.

Backed by the Rothschild family of European financiers, Rhodes bought a one-fifth share in the owners of the Kimberley Mine, the Kimberley Central Company, headed by a rags-to-riches cockney from London's East End, Barney Barnato.

Rhodes was soon able to buy out Barnato himself. A cheque for £5 338 650 changed hands, and in 1888 the Kimberley Mine was absorbed into the new De Beers Consolidated Mines Ltd.

Conditions in the diamond fields at the time were primitive and dangerous. Miners lived in tents, roughly-built 'homes' of wood and corrugated iron, or simply huddled beneath tarpaulins. Sanitary services were non-existent and no-one seemed concerned about health and hygiene.

Meanwhile colossal fortunes were being made, and Kimberley was booming.

Illicit diamond-buying flourished and, in 1905, the journal *SA Handbook* commented: "One of the most awful sights in the Kimberley of those days was the departure of an open wagon laden with chained White men, singing a dolorous song, for the Cape Town breakwater". (A reference to the notorious Breakwater Jail, in Cape Town's dockland area).

The Kimberley Mine, or Big Hole, closed in August 1914, at the outbreak of World War I. In its 43 years of life it had yielded 25 million tons of soil and rock, and 14,5 million carats of diamonds valued at £47 139 842.

In 1917 Ernest Oppenheimer, the head of A Dunkelsbuhler & Company, one of the smaller of the diamond firms that made up the syndicate through which De Beers' production was marketed, founded the Anglo American Corporation of South Africa with an initial capital of £1 million. He became a director of De Beers in 1926, and in 1929 he realised a long-standing ambition when he was elected chairman of the company.

Under Oppenheimer's direction, and with the co-operation of the government, the Diamond Producers' Association was formed, to regulate the policy of the trade. And all the major producers of diamonds in Southern Africa were brought under the direct control of De Beers.

Collins adds a new facet to the hunt for gems

In 1961 Sam Collins opened a whole new chapter in the story of the search for diamonds, bringing an innovative approach to the mining of the gems off SWA and venturing further than the succession of prospectors who had been working the area for more than 50 years.

Diamonds had been part of the SWA story since soon after the turn of the century. But, although the occurrence of diamond deposits along the south-western coast of Africa had been known since 1908, it was not until the 1960s that earnest attempts were made to follow these deposits westward, into the sea.

And that is where Sam Collins and his associates now came into the picture.

The first diamonds to be found in the territory – then German South West Africa – were said to have been picked up by a labourer on a railway construction site between Luderitzbucht and Kolmanskuppe. The labourer dropped his spade and carried to his foreman, August Stauch, what he called "a pretty stone".

Stauch, a lonely railway employee at one of the most desolate and godforsaken outposts on earth, tested the stone on the glass of his wristwatch, and convinced himself this was a diamond. However, there was no-one available with the necessary expertise to confirm this.

PA Wagner, in his book *Diamond Fields of Southern Africa*, records this historic find thus:

"The discovery was actually made in April 1908, by a Cape 'boy', a former employee of the De Beers Company who, while working on the railway line in the vicinity of Kolmanskuppe, picked up several stones, which he at once recognised to be diamonds.

"The stones eventually got into the hands of a railway official of the name of Stauch who, having ascertained that exactly similar mineral crystals were abundantly scattered through a superficial deposit of a coarse, gritty sand, proceeded forthwith to peg off enormous areas of this material, thereby securing what is one of the most important holdings in the entire diamond field.

"Stauch's action did not at the time attract much notice, but about two months afterwards when the opinion of the Cape 'boy' was confirmed by Dr Range, the Government Geologist, there was a great rush on the part of the inhabitants of Luderitzort to participate in the wealth which had been literally spread out at their feet."

Once Stauch – at the time Inspector of Permanent Works on the Kolmanskuppe section of the railway – had established that the stones handed to him were diamonds, he went to Swakopmund to interview the landowners, the Deutsche Koloniale Gesellschaft.

The DKG issued Stauch six renewable monthly licences, each of which gave the prospector the right to peg a claim on one kilometre radius within a stated area, extending from Luderitzbucht to the Pomona northern boundary.

By August 1908 practically the whole of the ground north of Pomona had been pegged. Small syndicates or partnerships were formed, and these were closely followed by the formation of companies, nearly all of them holding a 50-year concession from the DKG.

The Sperrgebiet

The German government, in agreement with the DKG, decreed the desolate, depopulated coastal strip extending some 350 km north of the Orange River a restricted "Sperrgebiet", or diamond coast, in which to this day trespassers face prison sentences or heavy fines, with even heavier penalties if caught in possession of uncut diamonds.

Entry to the Sperrgebiet is by permit only – and permits have seldom been granted to people not employed by the mining authorities.

For nearly a century, tales of fabulous fortunes made from "pirate raids" on the Sperrgebiet have spread around the world, as the colourful language of fisherfolk in the area has magnified chance and scanty finds of diamonds, from time to time.

Unsuccessful attempts have been made to enter this lifeless, forbidden zone by sea and air. In one such attempt the pilot of a light aircraft landed on a remote pan, only to find a posse of police waiting for him, from a hidden security post!

Authors of best-selling novels have seen the notorious "Skeleton Coast" and man's quest for diamonds along this forbidding and forbidden shoreline of South West Africa (now Namibia) as an ideal subject for stories of action and adventure.

Geoffrey Jenkins, author of the popular novels *A Grue of Ice* and *A Twist of Sand* (the setting for which is the Skeleton Coast), followed these up with his book *The River of Diamonds,* published in 1964. The theme of this novel is a daring expedition to locate and recover the vast quantities of diamonds deposited on the sea-bed off the Sperrgebiet.

"The quest for diamonds," says the blurb on the dust-jacket of the book, "is one of the most dramatic, adventurous and exciting of man's endeavours. It has brought out the best and worst in him. The best because the searcher must pit his endurance and cunning against the forces of nature and overwhelming odds; and the worst because of the corrupting element in the prospect of great and sudden wealth."

Jenkins's gripping story, inspired by Sam Collins's quest for sea diamonds off the SWA coast, features a specially-built craft, the *Mazy Zed*, equipped with all the most advanced and ingenious technical apparatus.

The *Mazy Zed* is towed to the base of the operations, Mercury Island, offshore of the Namib Desert, only to be confronted by "a strangely magnetic Englishman with a mysterious past" who is determined to block the success of the operation and to destroy any trespassers on the island, of which he is headman.

Another internationally acclaimed South African author, Wilbur Smith, wrote a best-selling novel *The Diamond Hunters*, a story of "violent action, suspense and a background of the fascinating diamond world in Cape Town, London and the mines beneath the tumultuous seas off the West Coast of South Africa".

The theme of the novel is of an inheritance that brings its own curse, and of consuming jealousy and hatred within a family that work their way towards a climax of murderous destruction.

Guano islands off the German South West African coast, for so many years the subject of tales of mystery, mishap and intrigue, were first officially prospected for diamonds in 1910, when certain of them belonging to the Union of South Africa were visited by an expedition sent by the Union government.

The islands visited were Possession, Halifax, Penguin, Pomona, Ichaboe and Seal.

No actual prospecting work was done on some of these islands, but on Possession about 3 000 loads – or 16 cubic feet – of gravel were treated from various parts of the island and this produced 223,5 carats of diamonds, valued at £511 10s.

The biggest concentrations of diamondiferous gravel were found at the southern end of Possession, in a belt running from east to west and immediately above the high-water mark.

Because of the difficulties in working these deposits, and the great expense involved, it was decided to abandon Possession as a source of diamonds and to examine some of the other islands. These produced nothing during the 1910 expedition, and with total expenditure on the venture running to £825 7s 7d, it was wisely decided to leave the penguins undisturbed.

Enter Ernest Oppenheimer, and De Beers

From an early stage, De Beers Consolidated Mines Limited at Kimberley took an active interest in the fledgling diamond fields between Luderitzbucht and Bogenfels.

In 1914 Ernest Oppenheimer, then a 34-year-old merchant and an acknowledged expert on diamonds, and his associate Alpheus Williams visited German SWA at the request of De Beers to investigate diamond mining activities there, and to report back.

This they did and in a detailed 60-page report dated June 4, 1914, Oppenheimer and Williams were able to inform the chairman and directors of De Beers that control of the whole of the diamond fields of German SWA had been reduced to the two major companies, the Koloniale Bergbau Gesellschaft and the Deutsche Diamanten Gesellschaft.

They were further able to report that, soon after the discovery of diamonds in the territory, it had become apparent that the sales of

German SWA diamonds should be carried out through one channel, and it had been agreed that gems found should be disposed of through the government.

The necessary ordinance had been issued, giving the government the additional right to fix the quantity each company could bring into the market, irrespective of production.

The SWA government, the report said, could not attend to the sale of the diamonds itself, so the Diamond Regie had been formed, to last for one year. Its life had later been extended, and eventually the Regie had been succeeded by the Diamond Syndicate.

Following the overthrow of the German administration of the territory in World War I an offshoot of De Beers, Consolidated Diamond Mines of South-West Africa (known as CDM), was formed in 1919 and this took over the numerous small companies in existence at the time.

CDM secured exclusive rights for 50 years – since extended – over the entire region known as Diamond Area No 1, north of the Orange River.

Diamond production was restricted in the depressed times of the 1920s. Prospecting continued, however, and in February 1927, only a few months after the first major discovery of diamonds in the area, the government stepped in, securing vast areas to which the public was forbidden, and bringing mining operations under control so that the market would not be flooded.

Prospecting activities by CDM in 1927 revealed the presence of long stretches of ancient shingle beaches beneath the sand just north of the Orange. But it was not until 1945, at the end of World War II, that these old beaches proved to be immensely rich in diamonds. They have, ever since, been the object of CDM's mining efforts on a large scale.

Diamonds are found in a layer of gravel covered by a blanket of sand which in places is over 30 metres deep. Beneath the gravel is solid bedrock, pitted with crevices and gullies in which rich pockets of diamonds are also sometimes found.

About 30 km north of Oranjemund there is a sign that reads "King Canute Road" – so named because the road leads to the spot where men successfully, although temporarily, turned back the tide to recover the most sought-after gems in the world – diamonds.

Farther south, diamonds had been discovered in Namaqualand from as early as 1925, and to this day several companies continue to mine along the coastal strip between the Olifants and Orange rivers.

In June 1926, diamonds were first found in quantity in Namaqualand while Captain Jack Carstens, an officer in the Indian Army, was spending his leave visiting his father, a trader at Port Nolloth.

Jack Carstens stumbled by accident on a huge deposit of alluvial diamonds, news of which started a rush of prospectors who were prepared to face the hardships of life in this dry, hot and waterless area in their quest for riches.

Port Nolloth was jammed with prospectors who threatened to seize the diamondiferous areas by storm. The "rebellion" subsided only when police reinforcements arrived, and machine-guns were mounted in and around the desert port.

Hans Merensky's fabulous discovery

In spite of the difficulties, some fantastic finds were made. Dr Hans Merensky, a renowned geologist whose name became linked with the prospecting of platinum at Rustenburg and the mineral riches of Phalaborwa, picked up 487 diamonds from under one flat stone, and he recovered a total of 2 762 diamonds in the month of September 1926 in the Alexander Bay area.

Merensky and other geologists observed a baffling phenomenon: the diamonds were invariably found in beds of gravel mixed with the fossilised shells of an extinct warm-water oyster known as *ostrea prismatica*.

Where the diamonds came from remained a mystery. For long it was thought they had been washed down the Orange River, and there were legends of lost valleys of diamonds. But prospectors who searched the bed of the river found nothing to substantiate this.

Likewise, the origin of the undersea gems found on the Atlantic coast from Angola down to the Olifants River is still a mystery. A number of theories have been advanced over the years – one of them that the diamonds were carried to the sea by glaciers from the heart of Africa during the ice age.

Lawrence Green, in his book *To the River's End*, said that at first the discovery of diamonds along the German West African coast was taken as strong evidence in support of the gems having a marine origin.

"The diamonds on this coast are of a type entirely different from those of Kimberley, and the South African river diggings. They are smaller and more brilliant, even when uncut. Their purity is remarkable.

'German geologists who investigated this sudden windfall in the hitherto barren colony in 1908 put forward another possible explanation of the presence of diamonds. The southern boundary of the colony was formed by the Orange River, and it was suggested that the diamonds could have been swept down to the sea by that great stream, and then carried north along the coast by the Benguela Current, and distributed while the land was still submerged.

"This theory received strong support in 1927, when the great treasure chest of diamonds was opened at Alexander Bay, just south of the Orange. It was simple to deduce that a similar deposit should be found on the north bank.

"Close to the pits sunk nearly 20 years previously, the Orange Mouth diamond terraces were located. Ever since then there have been busy diamond workings within sight of each other on the north and south banks of the Orange."

There were still geologists, Lawrence Green wrote, who believed that some of the diamonds originated in a submerged "parent rock" off the coast, and that they were carried along a "pipe" to the shore.

"In the early days the German government accepted this theory and by Imperial decree vested all rights in diamonds on the sea floor in the colonial treasury."

Green noted that while geologists still avoided being dogmatic about the origin of the South West African coastal diamonds, "it is no longer the deep mystery it was at the time of the first discovery".

Experts' theories on the origin of sea diamonds

A General Mining metallurgist, Bill Webb, stated in a paper on sea gems that diamonds were formed naturally in a type of lava, and were found in volcanic kimberlite pipes which thrust their way to the surface toward the end of the Cretaceous period, when extensive volcanic action was scarring the rising land mass of Southern Africa.

"Kimberlite," he wrote, "is a complex igneous material: a hard but chemically unstable rock that disintegrates through weathering. The wind, the sun and the rain reduce the green rock to yellow clay which is washed away, releasing its hard crystalline components.

"Garnets and diamonds are set free to wander the face of the land or, because of their high specific gravity – some three-and-a-half times as heavy as water – to settle in deep pockets and lie hidden by layers of lighter material or, if not checked in this way, to be washed on and into the sea.

"When diamonds were first discovered in South West Africa, their origin there was even more of a mystery. They were found in desert valleys, loose in the sand where native labourers picked them up by the thousands, crawling on hands and knees in line abreast down the valley and popping the precious stones into cans slung from their necks.

"Nowhere was there evidence of kimberlite pipes.

"At first it was supposed that the diamonds and garnets had weathered from the granites and pegmatites of the district, until some geologists concluded that the diamonds may have come from the sea near Pomona, a bird island south of the teeming seal haven of Albatross Rock."

Bill Webb said mining men favoured the view that larger diamonds were found high in kimberlite pipes, "and although there is no definite geological evidence to support this theory, it is interesting to note that the alluvial deposits, akin to the one in which the Jonker diamond was found, some 5 000 feet above sea-level and as far as 600 miles inland from the Orange River mouth, yield a higher percentage of large stones when compared with marine terrace diamonds.

"If this belief is correct, the sea may still yield greater treasures in the 'missing' large stones from the upper reaches of long-eroded pipes."

Webb added that while the origin of diamonds being recovered from the sea remained a mystery, the evidence so far pointed to both a land and a sea origin.

Prof John H Wellington, at one time head of the Department of Geography at the University of the Witwatersrand, noted in a chapter on mineral deposits along the Southern African coast, in a work published in England in 1955, that:

"It is interesting to find that raised beaches on Possession Island are also diamondiferous, leading to the surmise that the origin of the diamonds is a pipe in the sea-floor".

The late ETS "Ted" Brown, who had a lifelong association with the diamond industry and who retired in 1976 as an executive director of Anglo American Corporation and a director of De Beers Consolidated Mines, was one of those who firmly believed coastal diamonds came from the sea, and not down-river.

Referring to areas prospected by Consolidated Diamond Mines of SWA and Tidal Diamond Corporation (TDC), lease-holders of the coastal strip between Hottentot Point north of Luderitz, and Sandwich Harbour south of Walvis Bay, he said:

'To my mind it is quite clear that the CDM and Tidal diamonds, which appear so different from the diamonds from pipes inland, have different origins. This means the pipes from which these coastal diamonds come are more than likely situated offshore, not inland.

"The problem of extracting diamonds from the sea does not strike me as being as difficult – depending of course on the water-depths – as getting oil from offshore. In the future this offshore area must be an important one for De Beers. Meanwhile, prospecting in deeper water is continuing."

General Mining's Francois Hoffman, who was to become Sam Collins's chief geologist (he later became chairman of the Rembrandt-linked Trans Hex Group), theorised that while basaltic and kimberlitic pipes formed part of a series of volcanic intrusions only a few miles inland of the South West African coast, "who can say that a similar condition does not exist in the sea?"

"I believe" Hoffman added, "that an extensive pool of diamondiferous gravel, perhaps thinly scattered, lies in the ocean and as the continent rises, the undertow sets it in motion, grinding away the associated minerals and concentrating the diamonds.

"These then move up towards the land, where they are trapped with other material and deposited on the shore, waiting for the next elevation of the land to leave a new terrace stranded above the waterline."

Whatever his beliefs were at the time, "Hoffie" was destined to play a major role in the search of the seabed that was about to begin – a search which, although slow and frustrating at first, proved ultimately to be one of the great success stories of gemstone mining.

A hard-hat diver reports to marine superintendent Billy Evans, a Texan, in the search for seabed diamonds off SWA.

CHAPTER THREE

Sammy buys Johann Vivier's offshore concession

THE Collins sea-diamonds story had its origins in, of all places, a doctor's consulting rooms at Vredenburg, near Saldanha on the Cape West Coast.

Dr Piet Neethling, at the time of writing the only surviving member of the original group that initiated the diamonds-from-the-sea venture, has recalled how it all started. And while his recollection might differ in certain details from the Collins version, it is equally fascinating.

"In the late 1940s, when I was still practising medicine in Vredenburg, a very dear friend of mine, AP du Preez, and I started a business in the fishing industry on the Cape West Coast.

"And as I recall it a man named Johann Vivier, a sheep farmer from Beaufort West, came to see me in my consulting rooms one day, in the late 1950s. He told me he had acquired a sub-concession for the exploration of diamonds off the coast of South West Africa, and couldn't we help by supplying a boat that could be used for recovering diamonds from the seabed.

"What had happened apparently was that a gap had been identified in the concessions already awarded, and he and his partners had gone to the SWA administration about it. This gap was in the area offshore of the landward concessions already awarded at Oranjemund. It extended seaward from a line between the high and low-water marks – what was then known as the Admiralty Strip. This offshore concession included some of the islands off the SWA coast.

"This became a tremendous story, the area involved being adjacent to and offshore of Anglo American's diamond mining concessions at Oranjemund.

"Vivier said he now had this sub-concession, and that if we could provide a boat with the necessary suction pumps and other mining equipment, we could go into this as a joint venture. He believed there were a lot of seabed diamonds in the offshore area between Oranjemund and Luderitz.

"So we got hold of a boat through someone else with an interest in this venture, had it rigged out with pumps and other diamond recovery equipment at Hout Bay and we hired a skipper and crew to sail it up to Luderitz, which seemed the obvious place to go.

"But when it got to Luderitz the port authority would not allow the boat, with all this weird-looking paraphernalia on board, to operate from there without the necessary permit. Eventually, after about a month of being laid up there without any change in the situation, the crew got fed-up, walked off the boat and returned to Cape Town by bus or train.

"What eventually happened to that boat I can't remember; it probably went back to where it had come from. But that was how the quest for sea diamonds really got started. This was a first, halting step towards what eventually happened."

Later, says Dr Neethling, after news had been received that the concession held by Vivier had begun to yield positive evidence of the presence of diamonds, he and his friend and business associate 'Aap' du Preez decided they should do something about this potentially rich source of gemstones.

"At that time we had a lot of dealings with the late Abe Bloomberg, a leading Cape Town attorney, parliamentarian and businessman, and a very clever person in whom we had the fullest confidence. We told him about the possibility of mining diamonds in the sea.

"Abe Bloomberg said he knew someone by the name of Peter Keeble who would be going to the United States soon and, perhaps, he might be able to put us onto something. Keeble in turn knew about Sam Collins, who was then running his Collins Submarine Pipeline Company in Texas, and as I understand it he phoned Collins from New York and told him about the possibilities of mining diamonds from the sea off South West Africa."

So, by the time Sam Collins became interested, about 1960, in tendering for the installation of a two-mile-long submarine pipeline to supply fuel direct from tankers to the CDM mine at Oranjemund, he was already convinced from what Keeble had told him that sizeable quantities of recoverable diamonds were to be found under the sea in the area.

While Collins Submarine Pipelines did not succeed in its bid for the Oranjemund pipeline contract, it could be said that this De Beers project, with its far-reaching effect on the onshore infrastructure, was a factor in triggering the start of the marine diamond industry off the SWA coast.

To Collins, as it was to those who had set him thinking about sea diamonds, it was only logical that if diamonds had been found in quantity along the shoreline – as he had observed on the beach diggings at Oranjemund – they must surely also be found in quantity on the adjacent seabed.

He came to South Africa for the first time in the early 1960s, in connection with the laying of an effluent pipeline out to sea for

Titan Products at Umbogintwini on the Natal South Coast. He was accompanied by his wife Georgette, a strikingly attractive Texan, widely known as Gigi. Her birthplace – by virtue of the fact that her mother happened to be in Scotland at the time – was Glasgow.

Tromp van Diggelen's memoirs grip Sammy

Socrates Vartsos, a mining geologist who worked closely with Sam Collins in his sea-diamond venture in the 1960s, recalls that during this first visit to Natal, Collins expressed keen interest in the memoirs of the late Tromp van Diggelen, which he had been advised to read.

Van Diggelen had for many years been a much-publicised South African "strongman" who had run courses in body-building and physical fitness. He was a contemporary and a disciple of the famous international "strongmen" of the time, Eugen Sandow and George Hackenschmidt.

The South African had also been involved in many off-beat private salvage ventures including attempts to recover treasures on land and at sea. He recounted these ventures in his memoirs, *Worthwhile Journey*, first published in 1955.

The part of this book that particularly interested and fascinated Sam Collins was that in which Van Diggelen confirmed the finding of diamonds off the mouth of the Orange River as early as the 1930s. What was more, he said he knew the exact location of these gems on the seabed.

Tromp van Diggelen was, in his later years, regarded by many as somewhat eccentric, so his autobiographic reference to diamonds in the sea off SWA was taken seriously by few people. Sammy Collins was one of the few. It seemed to him to lend credence to what he had already heard from Keeble, Bloomberg and others.

In a chapter headed "A Fortune Escapes Me", Van Diggelen referred to work done along the South African coast in 1937 by the Italian salvage vessel *Arpione*, following a trip he had made to see salvors in Genoa on behalf of the African Salvage Corporation, of which he was a director.

It turned out, ironically, that while the vessel made a handsome profit for its owners from a large quantity of copper recovered from a sunken wreck off East London, it had no similar luck on sites indicated to it by the ASC of wrecks of early Dutch vessels known to contain valuable cargoes.

The *Arpione* could remain on the South African coast only for a fixed period, and although much searching was done with heavy grabs, nothing of value was found. And Tromp van Diggelen recalled in his book that before the vessel headed up the West Coast on its return to Genoa:

"Now came the truly romantic side of the whole adventure. Baron Manzi Fé (of the Italian salvors, Societa Ricuperi Marittimi) and the ship's captain visited me in my office.

"In true schoolboy manner, we three drew up a contract wherein it was agreed that, in return for my showing them where diamonds could be found at the bottom of the Atlantic Ocean at the Orange River Mouth, and outside 'territorial' waters, I would receive 50 percent of all profits."

But, Van Diggelen added, it was the weather that robbed him this time of another much sought-after prize – "and they (the diamonds) remain at the bottom of the mighty ocean!

"Again a fortune was to be denied to me. It was quite evident that

It is in gullies and crevices like this that diamonds are found offshore as well as onshore, on the West Coast.

speculative money was never to be mine! I had missed the 'Kruger Millions', and my father had lost three great gold mines and the possibility of possessing the greatest diamond mine in all the world.

"I *knew* that diamonds were right underneath the salvage vessel as she steamed around for a whole, hopeless week off the Orange River mouth. Wretched weather and strong winds made it impossible to anchor securely enough to send down suction-hoses, divers or even heavy grabs.

"The weather allowed only light grabs to be used. These failed to penetrate the deposit deeply enough to reach the diamonds.

"The *Arpione* was under suspicion, and coastguard boats would cruise around at times. As she was outside the three-mile limit, however, nothing could be done to stop her carrying out her search with the small grabs.

"Even if diamonds had been found, they could legitimately be taken away as long as they were not landed on the coast of the Union, or of South West Africa.

"Eventually I took some bags of the lighter deposit brought up by the grabs to the experts at the University of Cape Town. I was shown that the stuff *did* contain what are called 'sand' diamonds – little things that were too small to wander down through the gravel. Under the microscope they showed the correct octohedron crystallisation always found in stones that are big enough to be cut, or to be used commercially in diamond drills and other instruments."

" The diamonds are still there", wrote Tromp

Tromp van Diggelen concluded this revealing chapter:

"The diamonds are still there, and I know the spot!

"In 1954, a better-equipped vessel than even the Italian – the famous British salvage ship *Twyford* – came to our waters and carried out successful salvage work on copper ships that had been sunk during the Second World War.

"I was sent for to interview the agents, the skipper and the sailing-master of the ship. I told them about my tin-ship and the Orange River diamonds. Later a representative of the salvage firm flew out from England.

"We two spent an hour poring over marine charts. The same day he flew back to Southampton, and nothing more has transpired.

"That speculative fortune still awaits me around some corner. The true British seadogs I had interviewed were impressed. Perhaps the authorities in England thought it was not quite *de rigueur* to look for diamonds which were 'just about' on the three-mile limit.

"What the amazingly well-equipped *Twyford* might have done with her powerful suction pumps, and the diamond washing equipment I had planned for her decks, is something to dream about.

"However, 'the stuff that dreams are made of' never *did* make anyone rich.

"Yes, I really *do* know where those diamonds are – and very exactly too.

"May they *not* rest in peace!"

Little did South Africa's one-time strongman know when he wrote his memoirs that they were to become a catalyst – a key to the door as it were – in the world's biggest and most-successful sea-diamond mining venture; a venture in which he would have loved dearly to be a participant!

Perhaps it was those emphatic last few lines of Van Diggelen's chapter, "A Fortune Escapes Me", that Sam Collins saw as a challenge, and which prompted him to take up where others had left off – and to succeed where they had failed.

Collins could also have been influenced in his decisions by the writings of Lawrence Green. Apart from the passages in his books already quoted, relating to sea diamonds, Green wrote specifically about the guano and "diamond" islands off the SWA coast in one of his earliest books, *Where Men Still Dream*, published in 1946.

He visited some of these islands in 1916, and he refers in one of the chapters of this book to Mercury Island – "a sea-swept rock with a most appropriate name. This black, oblong bird sanctuary has an enormous cavern in its face, and the island contains a maze of tunnels. When the seas rush in, the whole island shakes like quicksilver.

"Visitors are rare; but when they do make the difficult landing on Mercury Island they seldom feel safe for a moment.

"On the summit of Mercury, 130 feet above the sea, there is a funnel which they call the 'Glory Hole'. You can hear the noisy waters far below if you listen at the edge of the funnel. Undoubtedly it opens into the great east-to-west cavern.

"There are many stories of the 'Glory Hole', but no records of exploration. Diamonds are there, they say..."

"Possession: island of mystery – and diamonds..."

Lawrence Green also refers in this book to the 1910 haul of diamonds taken from Possession, "this graveyard of the seals".

"Is there another tiny island in the world which has produced diamonds?" he wrote. "I doubt it."

"The discovery of diamonds on Possession came about as a result of rich hauls which were being made on the mainland – at that time German territory. One of the theories of the origin of these beautiful stones was that they reached the shore from a 'parent rock' on the ocean bed.

"It seemed possible there might be diamonds on the British islands too, and the government officials and prospectors combed the bird sanctuaries for this new source of wealth.

"And on Possession, among dozens of worthless crystals, diamonds were found. A thorough search followed. Washing machines, sieves, picks, dynamite and spades were sent to the island.

"Ovambo native labourers with experience on the German SWA fields were recruited for the work, and a trench three-hundred yards long in a bank of gravel and clay, and the all-pervading seal hair, was dug."

Lawrence Green recorded that, after it became known that diamonds had been found on Possession but that the government venture had been abandoned because the results had not justified the expense, a syndicate of eager prospectors in Cape Town planned a daring scheme.

"With diamonds on the mainland and on the island, they argued, there must also be diamonds on the floor of the sea. They would dredge for them, and bring up a richer load than any fisherman had ever hauled!

'The coasting steamer *Nautilus* was chartered, with all the secrecy appropriate to such a venture, and the prospectors steamed north.

"I heard the tale from a man who had tried his luck on every African diamond field – Dave Wilson, a tough, sun-browned Scot who sailed in the *Nautilus*. 'We lowered the grab-buckets to test the bottom near Possession, inside the three-mile limit', he called wistfully. 'Misty weather it was, just what we had prayed for.

"A diamond came up in one of the first buckets. We did not need to test it; we old prospectors knew. And just as we were all crowding round in fine spirits, cheering and slapping each other on the back; just then the sharp bows of a gunboat poked out of the fog.

"There began a game of blind man's buff. The gunboat had three times our speed – but the fog was in our favour. A solid white wall shut down between the gunboat and the *Nautilus*. We should have steamed clean away, but our skipper had forgotten all about Possession Island. The shock as we struck the reef threw us all off our feet. All hands reached the island in the boats. Diamond diggers' luck, I suppose – I have known worse!'"

And there, Lawrence Green concludes this story, the *Nautilus* remains, "wedged out of reach of the gale-driven combers, a silent memorial to old adventure.

"The legends of hidden caches of diamonds on Possession are innumerable. I know of two illicit expeditions in search of hoards which were certainly not based on mere rumour."

Green says that in the graveyard on Possession, one parcel of diamonds was supposed to have been hidden – among the bleached white crosses of decaying wood where the bodies of the captain and his wife from the sailing ship *Auckland* were buried when they washed ashore there, after the ship was wrecked on the island with the loss of all hands, many years ago.

"I can think of many places where I would rather seek treasure at night than in the graveyard at Possession. They say that the *Auckland*'s captain and his wife may still be seen standing miserably on the beach, near the wreck. Sharks guard the island. The ghost of the woman, they say, has no legs...

"Such is Possession, island of mystery – and diamonds."

A deal is done – and Sammy is bitten by fleas!

It was against the background of stories such as these, which Sammy Collins had read or been told about, that he took a decision that was to make mining history.

In 1961 – the year that South Africa broke with the Commonwealth and declared itself an independent republic – he and Emerson Kailey successfully negotiated the purchase of the government offshore mining lease held by Johann Vivier and his partners, the Van Zyl brothers.

Gerrie van Zyl was a well-known West Coast industrialist and a driving-force behind Marine Products, and his brother Johannes became a senator in the SA Parliament.

Vivier had by 1961 produced conclusive evidence that there were gem diamonds not only in the wadeable shallows off the SWA shoreline, but also in deep water within the concession he held. He had in fact become the first person to suck diamonds – albeit a very small number – from the seabed.

As the Afrikaans journal *Tegniek* put it, over a photograph of Vivier in a special feature on sea diamonds: *Hy Was Eerste* (He Was First).

So Sam Collins and his associates were by no means going into this venture blindly. For them it was a matter of taking up where Johann

Much of the Namibian coastline looks like this: rough and rugged, often fog-bound – and altogether forbidding.

Vivier had left off – and proving that there were diamonds on the seabed in large enough quantities to justify a full-scale mining operation.

The purchase price of the lease they bought from Vivier and his partners was reported at the time to be R480 000. The deal was done in Cape Town, and Sammy recalled afterwards: "We stayed at a hotel in the city where we got fleas. I'll never forget them!"

Dr Piet Neethling takes up the story:

"We met Sam Collins in Abe Bloomberg's office in Cape Town. The others present were Aap du Preez, Johann Vivier and Peter Keeble. It was therefore these six people including myself who really got the sea diamond mining industry going in Southern Africa.

"It was at that meeting that Abe Bloomberg drew up the joint venture agreement and we formed ourselves into a company called Marine Diamond Corporation, with Sammy Collins as chairman and managing director."

At the time that Marine Diamonds came into being Collins's pipeline company had heavy liabilities because of unsettled contract payments elsewhere in the world, so he had to seek financial backing locally for the costly operation he was about to mount, in the hunt for sea diamonds.

He succeeded in doing so through Colonel Jack Scott of General Mining who, on an overseas visit, managed to persuade the Standard Bank of South Africa in London to advance the money needed after he had told the bank he had the utmost faith in the diamonds-from-the-sea venture.

Collins later said that if it had not been for Scott, sea diamonds would never have become the source of wealth they ultimately proved to be.

"No-one was more enthusiastic about the diamonds-from-the-sea venture than Jack Scott. His confidence never wavered from the beginning. All the people at De Beers tried to discourage him, and actually there were very few, if any, besides Scott who believed diamonds could be taken out of the sea economically – if at all."

In London, Collins and Kailey had found that potential backers of their venture had pulled out on learning that it did not have the support of Anglo American and De Beers.

Later, however, the chairman of these giant corporations, Harry Oppenheimer, approved of General Mining participating in the scheme.

And Jack Scott managed to persuade SG "Slip" Menell of Anglovaal to join General Mining as a shareholding partner in the venture.

Sea diamond hunters described as "crazy"

This all happened against a background of scorn and derision that had been poured on the efforts of the earlier hunters of sea diamonds.

In 1958, when Johann Vivier and his associates acquired this first concession for the mining of diamonds on the seabed they were mocked and laughed at, and described as "crazy" by fellow prospectors and others.

At that stage, there were few people around who believed that diamonds could be recovered from the sea in payable quantities. "All the experts," Collins later recalled, "advised us against it as they thought it would cost more to get the diamonds out than they were worth."

Socrates Vartsos remembers being shown various reports issued by Anglo American, and signed by the corporation's chief geologist at the time, stating that if there were diamonds in the sea it would be uneconomical to mine them.

Vartsos, who had been looking for diamonds at the bottom of river systems in the West African state of Sierra Leone, says he disagreed with Anglo American's conclusions and accepted an offer to join a team of specialists in the Collins venture, which he felt was well-motivated.

The Vivier-Van Zyl partnership had meanwhile been working their offshore concession by using labourers who scooped up gravel in the shallows and sorted it on shore. Small quantities of diamonds had been found this way.

The concession area lies between the Orange River mouth and Diaz Point near Luderitz, some 320 km up the coast, and is located between the high-water mark and the 200 metre isobath (water-depth contour).

Johann Vivier later recalled that he and his labourers were under constant surveillance by diamond security detectives as they worked in the ice-cold surf, about 80 km north of Oranjemund and immediately adjoining the CDM onshore concession.

One day, in 1958, 39 diamonds weighing a total of 21,5 carats were found when gravel collected in a petrol-drum was tipped and sifted at the makeshift camp ashore.

"This was the greatest moment of my life", said Vivier. "I was terribly excited, and when I took the diamonds to show Mr Davis, the manager of CDM, he was astounded because these were the first diamonds actually taken from the sea."

Encouraged by this find, Johann Vivier bought a 50-foot fishing boat, the *Karibib*, at Walvis Bay. He also hired a deep-sea diver to take part in his first offshore venture to determine the whereabouts of diamondiferous gravel on the ocean floor.

Seventy-two-year-old Johnny de Olim of Cape Town, a fishing boat owner and former skipper who grew up in Madeira, takes up the story.

"Johann was advised to hire a Portuguese skipper for his boat, because we Portuguese knew that coast better than anyone. I was working at Luderitz at the time, after sailing my boat there from Madeira in 1951, and he approached me there. After telling me he was looking for diamonds, he offered me £100 a month and five percent commission on any diamonds found. This was an attractive offer, and I took the job.

"Vivier, who owned some farms in South West Africa, had his own plane – a four-seater – and he flew me from Luderitz to Walvis Bay to become skipper of the *Karibib*. The boat was fitted with pumps, suction pipes and other gear, and after calling at Luderitz, Johann told me to head for the Orange River mouth where he believed we would find lots of diamonds.

"After working there for a while without finding anything, I asked Johann exactly what he was looking for. He said gravel. I asked him why he hadn't said this in the first place, because as a fisherman I knew where the gravel was.

"On my directions we headed northward again, towards Luderitz. I took the boat straight to where, from experience, I knew pockets of seabed gravel to be.

"The diver, a chap we called 'Pikkie', wearing old-fashioned diving gear – copper helmet and all – went down and placed markers, with buoys attached, on the gravel-bearing areas. This was close inshore, about 20 miles (32 km) south of Chameis Bay. We were working in about 13 fathoms (23,7 metres) of water.

"We brought up 12 bags of gravel in that area, and on sifting it we found 10 small diamonds in the first bagful taken. These then became the very first diamonds taken from the deep sea, as distinct from those found in the surf, along the shoreline.

"The diamonds were taken to Oranjemund, and that was that. I was paid my wages, plus £10 as commission on the diamonds found – and I returned to my job as a fishing-boat skipper at Luderitz.

"Johann didn't take it any further. He told me he was not interested in mining diamonds. All he wanted was to prove that there were diamonds on the seabed, and then to sell out. And that was how he became a very rich man."

"It was all very rough-and-ready..."

Mrs Babs Vivier, Johann's widow, recalls that her late husband had been prospecting for diamonds along the SWA coast for about four years before he and the Van Zyl brothers sold their offshore concession to Sammy Collins.

"The first gravel that Johann took from the seabed was collected in empty 44-gallon petrol drums, and then taken ashore for sifting, in hand-held gear. It was all very rough-and-ready."

During the interview with Babs Vivier her close friend from Beaufort West days, Mrs Valerie Dannhauser, widow of Dr Piet Dannhauser of Trovato Estate, Cape Town, interposed to say:

"I remember Johann telling us, with a laugh, that he didn't make money out of diamonds; he made it out of what he called his nuisance-value. When he first got a concession to mine diamonds between the low and high-water marks at Oranjemund, his scrapers would go onto the beach every day and push the sand up to the high-water mark, which was the outer limit of the De Beers concession.

"Eventually this created such a nuisance to De Beers, with Johann coming and going in their 'Sperrgebiet' with his workers and equipment, that they were only too pleased when he sold the concession. He got a lot more money out of this sale than he ever did out of pulling diamonds from the sea!"

The story of Johann Vivier, who died a millionaire in Windhoek in 1981, is a romantic one; a rags-to-riches tale of a determined and motivated man who, like Sam Collins after him, defied the critics – and the daunting elements – to prove that his beliefs were right.

Johann was born on the farm Boesmanskop in the Beaufort West district in 1906. When his widowed mother Johanna died, leaving no money, he had to quit school after Standard 8 and with only 17 shillings and sixpence in his pocket, he trekked to Cape Town. There, he got a job in a garage, at a wage of £4 10s a week.

He later moved to South West Africa where he was given a job in the government garage. He saved enough money to buy a remote farm in a particularly dry and desolate part of SWA, for only one shilling a hectare, and he himself built a 28-km access road to the farm, and drilled a borehole for water.

It was while he was recovering in hospital from an operation that he began pondering the possibilities of finding and extracting diamonds on the seabed off SWA, and after his discharge he went about trying to obtain a prospecting licence.

He was able to convince his friend Johannes van Zyl that such a move would be well worthwhile. Van Zyl duly applied for and was granted a mining licence in the name of a newly-formed company, SWA Prospekteerders, with himself, his brother Gerrie and Johann Vivier as equal-share partners.

The company, launched in 1956 with a capital of R84 000, was two years later granted an offshore concession by the SWA Administration in which to prospect and mine the waters adjacent to the CDM workings at Oranjemund. This was the concession later acquired by Sam Collins and his associates.

Johann Vivier could not have wished for a better outcome of his "crazy" notion that the seabed off SWA held a treasure-trove, waiting to be tapped. A man with a similar background to that of Collins, he also shared some basic character traits with him. What he lacked in the way of formal education he made up for in acumen and energy, vision and drive.

Like Collins, he met the right people at the right time, took his opportunities when he saw them – and never looked back.

Birth of the Marine Diamond Corporation

The R10 million Marine Diamond Corporation (MDC), founded in 1961 at the time of the purchase from Vivier and his associates, was to become the operating arm of Collins's Sea Diamond Corporation Ltd, the principal shareholder in the new venture.

The MDC was formed in a partnership with General Mining, Anglovaal and the Abe Bloomberg/Kappie Strydom group, later trading as Diamond Royalties and Holdings Ltd.

While Collins, through the wholly-owned Sea Diamond Corporation, acquired control with 40.9% of the shares, General Mining and Anglovaal obtained a joint holding of another 40.9% in the venture.

The Bloomberg/Strydom group received an option to purchase at par 12.5% of the shares of MDC, plus a royalty of 5% on the new corporation's gross production. The remaining shares – 5.7% – were acquired by the Vivier/Van Zyl group.

Sam Collins became chairman of both Sea Diamonds and Marine Diamonds. With him on the board of Marine Diamonds, of which he was also MD, were Emerson Kailey, Jack Scott, Kappie Strydom, Abe Bloomberg, AP du Preez, HN Hart, Peter Keeble, "Slip" Menell and Dr Piet Neethling.

Directors of Sea Diamond Corporation, apart from Collins, Kailey and Keeble, were A Webster, MP, D and J Ipp, J Doms and JW Mitchell and FL Martens, both British.

A lump of conglomerate taken from the seabed off Namibia, showing several small diamonds embedded in it.

CHAPTER FOUR

Mutiny at Luderitz: the 'Emerson K' saga

THE fledgling Marine Diamond Corporation, so it was reported at the time, proposed mining its newly acquired offshore concession by adapting the oldest and most modern metallurgical techniques to overcome the problems of mining diamonds in the open sea.

Marine suction pumps would be driven into the seabed gravel from floating plant, and the gravel would be treated either from floating dredgers or pumped ashore for washing and sorting.

With this in view, Sam Collins acquired a former British Admiralty salvage tug, HMS *Marauder*, and renamed it *Emerson K* in honour of his trusted partner and confidant, Emerson Kailey.

Kailey, who negotiated the purchase of the tug, was said to have paid the equivalent of R56 000 for it, and Collins then spent another R460 000 having it adapted for his planned diamond-mining venture. The 750-ton steam tug sailed from Southampton in mid-1961 under the command of a former mailship master, Captain DD Williams, and headed for the South West African port of Luderitz.

Geologist Francois Hoffman and metallurgist Bill Webb, who had been seconded to MDC by General Mining, met the tug when it arrived at the desert port. Sam Collins also arrived in Luderitz to see his latest acquisition and to take part in the sea-diamond hunt.

Webb recalls, with a chuckle: "Sammy invited Hoffie and myself to join him in a drink at a local hotel. I was surprised to see someone drinking whisky before lunch, and when we joined him he said with a laugh: 'Now I've got you fixed; I'm gonna report you both for drinking on duty!'"

Hoffman takes up the story:

"Collins wanted to get into prospecting for sea diamonds right away. He was a man in a hurry, and he had special recovery plant put on board the *Emerson K* there and then, in Luderitz. This turned out to be a total disaster. The plant just wouldn't work the way they installed it.

"It was built over the engine-room and there were awful spillages, with water going down into the engine-room and taking lots of sand with it.

Collins bashed by a mutinous crewman

"Except for the master and one or two others, the British-based crew mutinied, claiming they had signed on for the delivery voyage to Cape Town only, and not for prospecting for diamonds on the way. One of them hit Collins on the head with a blunt object, and Sammy reacted by pulling his gun. Then the police got involved, as it was illegal in South African law to point a firearm at another person except in very special circumstances."

(At the time South West Africa, as a League of Nations mandated territory, came under South Africa's jurisdiction).

The proposed early start to diamond prospecting had to be abandoned, and on the tug's delivery to Cape Town Peter Keeble paid off all the crew except the chief engineer, a Scot, the bosun – a cockney called Sibley – and the chief steward, also an Englishman.

When Collins returned to Cape Town he was amazed to see these three still on board, and demanded to know "who the hell kept these men on? – these were the goddam ringleaders of the mutiny!"

While the three ex-mutineers were not fired on the spot, they were warned by Collins that at the slightest sign of indiscipline they would be dismissed immediately.

Bill Webb recalls that in recruiting a new crew for the *Emerson K* "we had a hell of a job getting certificated mariners with deep-sea experience to accept that they would have to take the tug very close inshore during prospecting operations. In their training and experience it had been drummed into them that it was prudent practice to keep well clear of shallows, shoals – and the shore."

In Cape Town, the tug was properly prepared for its new role. It was equipped with an airlift plant, vibrating screens and the Pleitz jig – a tool developed early in the century by Albert Pleitz of Luderitz, for recovering diamonds from desert sands by eliminating the

Another lump of conglomerate, with a larger-than-usual diamond (about 2 carats) forming part of it.

lighter materials and concentrating the heavy minerals and diamonds, if any, for hand-sorting.

This jig, once favoured because of its water conserving properties in the parched desert, was about to be worked successfully surrounded by limitless supplies of water.

"We had to make a start somewhere..."

With its locally recruited crew under Captain George Foulis of Durban, the *Emerson K* sailed back, in October 1961, to the prospecting areas off the West Coast.

Captain Foulis, a former Safmarine officer, had been master of coasters operating between Durban and Mozambique. He joined Marine Diamond Corporation on the recommendation of Hout Bay millionaire Bill Mitchell, whose motor yacht *Schipa* had been delivered by Foulis from Durban to Cape Town some time before the formation of the MDC.

"Soon after the *Emerson K* had arrived in Cape Town I received an urgent telephone call in Durban from Bill Mitchell, asking me to come to Cape Town immediately as my services were needed by Sammy Collins.

"I managed to get a relief master for the coaster I had been commanding, the *Congella*, and I flew to Cape Town on a Friday. I saw Collins and the *Emerson K* the same day and joined the vessel as master on Saturday, to prepare it for sailing on the Monday."

Francois Hoffman, who was again one of those who sailed in the *Emerson K*, takes up the diamond-hunt story:

"So now, with the right sort of equipment, we had to make a start somewhere, and decide whether to search for diamonds originating in the sea or only for those carried there from land erosion.

"Being a geologist, I was expected to find the right places to look for diamonds on the seabed, but I had to tell them I really didn't know; that all we could do was to work on a logical basis, step by step.

"I suggested we steam as close inshore as possible, and look at the pictures formed by the depth recorder, and from these try and pick up gullies, crevices and so on. I also suggested we work in the vicinity of known onshore diamond deposits, moving about half a mile at a time and taking samples as we went along. In this way we could build up a picture of what the devil was happening at sea.

"On land, we had to have a crew to help us do a positional fix of our vessel so that we could plot with reasonable accuracy where we were, on the chart. It was slow-moving work."

The positioning of vessels engaged in the search for sea diamonds called for a navigational accuracy never before required at sea. This was done by using a land-based instrument known as the Hydrodist which, through a ship-to-shore reflection of radio beams, enabled a vessel to know its position to within one metre.

The Hydrodist is a modified, "marinised" version of the South African-invented Tellurometer, a sophisticated instrument for measuring distances with electronic precision.

When the MDC's mining operations were concentrated in and around Chameis Bay, a prefabricated centre for the land party was established on a hilltop on the northern limb of the bay – and this came to be known as Fort Reef.

At the time that the *Emerson K* sailed north-west from Cape Town, says Hoffman, a marine concession south of the Orange River had been registered in the name of a new prospecting company in the Du Preez/Neethling/Kappie Strydom stable – Atlantiese Diamant Korporasie.

"Du Preez and his associates had applied for this concession after Johann Vivier had been awarded the first offshore concession, north of the river.

"So, when the *Emerson K* sailed up the West Coast it first went into this southern concession before moving north of the Orange into the newly-acquired Marine Diamonds concession area.

"That first trip, of three weeks or more, proved abortive. Not a single diamond did we find. When the time came to return to Cape Town to refuel and replenish our supplies, we had only a few hours in port and then we were off again. Our families had to come down to the docks just to say hello and goodbye.

"On this second trip we again found no diamonds. Sam Collins was

On board Emerson K *in Table Bay, Collins congratulates his GM, Bill Webb, on the tug's first haul of sea diamonds.*

on board for these first few trips, and so was I, and it was difficult trying to conceal our bitter disappointment. We were spending large sums of money with nothing whatsoever to show for it."

A tough first year – with the critics gloating

The first year of the sea-diamonds venture was, recalls Francois Hoffman, "a truly tough and testing one for us.

"For months we went on searching without finding a thing. We even tried using small fishing boats, to get closer inshore, but still no diamonds – not even offshore of where Johann Vivier had found diamonds, practically in the surf."

(At one stage two small fishing boats, the *Yolanda* and the *Rijger*, were chartered and fitted out with small airlifts, trommels and jigs to test waters too shallow for *Emerson K*).

George Foulis says he soon came to respect Francois Hoffman as "an astute and logical geologist, having studied ancient as well as modern maps and sea charts, and having an impressive background of prospecting for mining companies.

"With Sammy and Bill Webb he worked out where diamonds were most likely to be found, and these are the places we went to – places such as the mouths of old river-beds, where marine diamondiferous terraces ashore had been broached and obviously washed into the sea.

"We also looked at places where reefs or shoals formed natural traps and we even went to places where old tailings, which we knew had not been worked efficiently, had been dumped into the sea. But none of these places yielded diamonds."

Captain Foulis says on one occasion, when he had Collins on board the *Emerson K*, Sammy wanted to go into Alexander Bay to do some prospecting. "I said we couldn't go in there; that the vessel would be endangered, with only a couple of feet under the keel.

"But Sammy was insistent. He asked if I was scared to go in there. I told him I was not scared, but that as master of the vessel, I was responsible for its safety. I then told him that I would act on his instructions only if he was prepared to sign a paper stating that, for mining purposes, he needed to go into Alexander Bay and that he would take full responsibility for any damage done to the ship.

"He agreed, and he gave me a signed note. We went into the bay, and I told Sammy he must tell me when to stop. Alexander Bay is a narrow inlet, and as we went in I could see that he was getting more and more worried.

"We got to a point where we had a mere 12 inches under the keel – and then he said that was far enough. As we backed out there was a shuddering, and I knew we had hit something. I checked, and was told it was the port propeller that had fouled a rock.

"I screwed the vessel over to starboard, and when we finally cleared that rock and got to the entrance I said: 'Okay Sammy, if you want to, we can mine here!' He replied: 'No, no, George, let's get a bit farther out'.

"In this way Sammy had tested me, and I had made my point. This was something I just had to do – and we understood each other a lot better after that.

"The port propellor was badly damaged, but the amazing thing was that it remained perfectly balanced, and caused no vibration in the vessel right up to when it was replaced some time later."

After the MDC's initial "dry-run" it began to look as though the sceptics who had scoffed at Collins's grand plan to recover diamonds from the sea were right after all. They now took every opportunity to gloat and to point out that the Collins scheme had been a "bum steer" right from the start.

At last, a sudden "turn of the tide"

Then, towards the end of 1961, the tide suddenly started turning in Sammy's favour.

"Our first break," says Francois Hoffman, "came when we were working off a vlei where there is very heavy sediment in the shallows, offshore. It was a beautifully calm day, and we were able to penetrate to a good depth. Our air-lift pipe just went down, down, down, to 40 or 50 feet, and our divers reported that the end of the pipe was well into the seabed.

"We hit gravel, which came onto our jigs – and there we found our

A close-up of the first diamonds recovered by the MDC from the seabed off SWA – with a sixpenny piece for size.

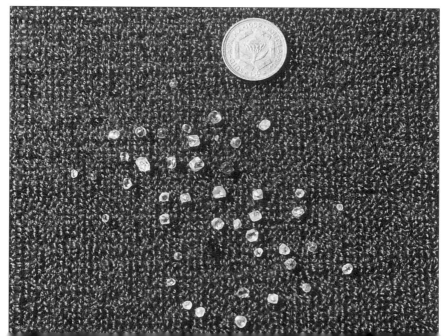

first three diamonds. Very nice gems they were, too.

"We then moved farther north, toward Luderitz, and while doing what we had been doing all along, and taking regular samplings, we hit it in a big way. We brought up something like 38 diamonds; typical West Coast offshore diamonds, small, clear, very cuttable and of a very good quality all round."

George Foulis's recollection is that after he had taken the *Emerson K* into a succession of inlets and bays and other "promising" places along the SWA coast, without result, a new strategy was adopted. It was decided to work down the coast in shallow water from Diaz Point, Luderitz, taking samplings about every 500 metres, until diamonds or diamondiferous gravel were found.

"We did this, and it was not until we got to a place called Wolf Bay, south of Luderitz, that we found our first diamonds in quantity. I recall that there were 13 of them in the first lot we brought to the surface. We went on to find extensive beds of diamondiferous gravel at Chameis Bay, Baker Bay and other areas."

Sea Diamond Corporation Ltd later placed on public record, in London, that the first diamonds in any quantity were discovered by the *Emerson K* in Wolf Bay on November 15, 1961, and that the tug returned to Cape Town on December 9 with 45 gem diamonds weighing a total of 8,96 carats.

The company, in a statement signed by Sam Collins, reported further that:

"By the end of March 1962, the *Emerson K* had sampled on a regional scale along the coastline from the Olifants River to Diaz Point (Luderitz) and altogether 44 more diamonds weighing 11,70 carats were recovered.

"Diamonds were also recovered from a few sites along the Consolidated Diamond Mines coastline and promising results from the Chameis area between Chameis Head and North Rock were recorded on April 16, 1962, when a total of 38 diamonds weighing 14 carats were recovered from a working site.

"Further prospecting in that area proved that an extensive deposit carrying diamonds was in existence. By the end of June 1962, all operations in the South West Concession had yielded 179 diamonds weighing a total of 58,81 carats."

Sammy elated at the good news

Francois Hoffman remembers the *Emerson K*'s first big haul of diamonds, after a disappointing and depressing start, as being "a really big story".

"Sammy Collins wasn't with us at the time; he was in Cape Town. We couldn't call him by radio because others would be able to hear, so we called our land party by walkie-talkie and got them to motor down to Oranjemund to make the telephone call that would tell Sammy we had made it.

"He of course was elated at the news – and this is how the mining of sea diamonds off the SWA coast eventually got going."

It was about this time, says Hoffman, that MDC also started planning geophysical sea-bottom and sub-bottom surveys, to help produce a better picture of the ocean floor.

"In this process of working, we then had our first results at Chameis Bay, between Luderitz and Oranjemund, and at North Rock and South Rock in the same area.

"These were very important discoveries, because not long afterwards they were able to sustain our mining operations, using specially adapted recovery barges."

Collins was quoted in Cape Town as saying the *Emerson K* had found diamonds worth about R120 000 off the West Coast. He also said the Marine Diamond Corporation had discovered nine potential offshore diamond mines in the area, which he claimed were as rich as the De Beers mines at Oranjemund – the richest diamond mines in the world.

The Texan was also reported to have decided to sell his diamonds through De Beers's Central Selling Organisation and that in anticipation of this, people were already buying the De Beers shares which had increased in value.

Sam Collins's calculated gamble was about to pay off.

Prominent parliamentarians involved in the search for sea diamonds: Kappie Strydom (left) and Jan Haak (centre).

Sammy Collins visits his first floating mining unit, Barge 77, *off the coast of South West Africa in 1962.*

En route to SWA, Barge 77, *the first mining unit, probes the seabed near Olifants River Mouth on the Cape West Coast.*

CHAPTER FIVE

Barge 77 leads the way and confounds the critics

"Certainly, Mr Collins's work in recovering diamonds in these novel conditions is a technical achievement of a high order."

— Harry Oppenheimer, March 1963.

THE *Emerson K*, having succeeded in her exploratory role, now clearly had to make way for a vessel or vessels more suited to mining operations on a large scale.

In 1962 a flat-bottomed barge used in Collins's pipeline operations in the Mexican and Persian gulfs was adapted to become the forerunner of a fleet of floating mining camps, equipped for the recovery of diamonds from the seabed.

This pioneering barge, registered as *CCC 77* but better-known as *Barge 77*, was fitted with a 30-ton-an-hour recovery plant and accommodation for 53 men. It could produce its own fresh water, had a fully-equipped laundry and a cold-storage capacity and stores to last three or four months.

The barge's radio equipment allowed direct communication with the home base in Cape Town, through specially developed single-sideband sets produced by MDC's technicians to overcome the West Coast's notorious radio communication defects.

While not self-propelled, *Barge 77* was able to move over an area of more than 1-million square metres within the zone of its four anchors, controlled by a specialised anchor winch.

Mining was done by dredging with a 30-cm (12-inch) airlift designed by Collins, using control methods developed during operations to keep it working in the sea in all its moods.

Silt, sand, gravel and boulders the size of a man's head could be sucked up with tons of seawater, and delivered to a scalping screen which eliminated the larger fraction in which no diamonds occurred.

Barge 77, towed into position in the MDC concession area by the *Emerson K*, started mining in August 1962, and in its first 11 months of operation, 51 000 carats were recovered.

By mid-1963 this ugly and ungainly-looking floating platform with its maze of steel pipes, valves, winches and mining paraphernalia had not only proved the existence in quantity of gem diamonds on the seabed; it had also started in earnest a new-type mining industry for which the future looked bright and promising.

The venturesome Sammy Collins had played his hand and won. He had confounded his critics and won the renewed confidence of those who had supported him from the outset.

An enlarged and mounted photograph in the possession of Mrs Babs Vivier shows a glittering "pancake" of diamonds taken from the sea, and it is captioned: "First 5 days' production, 2 100 stones, 1 018 carats."

A handwritten note under the photograph reads: "To Johann Vivier, September 12, 1962. Hope you get more and bigger ones, Johann, you really deserve it! Sincerely, Sammy Collins."

(In 1962 this particular haul of diamonds would have been worth less than R30 000. At the time of writing (1996), with the market price around 200 US dollars a carat, it would be worth about $204 000 – or nearly R1-million).

Prospecting results by late 1962 were considered sufficiently encouraging for the registration in South Africa of the Southern Diamond Corporation (SDC), which was awarded marine prospecting and mining rights along the coast between the Olifants River and the Orange River.

Lavish Texan-style parties became commonplace in Cape Town and it soon became obvious that Collins, flush with the early success of his venture, had established good connections in the right political, social and business circles.

In a major article on Collins and Marine Diamonds in the May 1963 issue of *Shell in Industry* it was said that:

"A fabulous legend has become reality almost overnight: the proved presence in the seabed off the desert coast of South West Africa of countless millions of carats of diamonds.

"They are there almost literally for the picking.

"An engineless barge, specially equipped to suck the diamond-bearing aggregate from the seabed, has already achieved sensational results in what has become the largest prospecting and mining concession in the world.

"It represents a mere pilot plant, and when it is replaced next year by a much bigger vessel, modified and equipped to bring to the surface by pipeline far greater quantities of 'pay dirt', the output of diamonds – almost all of them pure gemstones – will increase enormously.

"Old-time prospectors who have handed down over the years the legend of untold riches on the seabed off South West Africa would gasp in incredulous amazement if they could see the glittering prize

Barge 77 *encounters rougher seas and generally hostile conditions south of SWA's notorious Skeleton Coast.*

of a single day's operations offshore by the barge now moored near the drab coast.

"Sammy Collins has poured in hundreds of thousands of rands of his own capital – and is continuing to do so as startling evidence of the presence of thousands upon thousands of carats of pure gemstones make it increasingly clear that his bold near-gamble has paid off."

The *Shell in Industry* article said the popular belief was that Texans were always prepared to take a chance, and that they seldom did things by halves.

"Sammy Collins is the embodiment of this 'Lone Star' tradition. Yet he is no gambler in the careless sense of the word. When he heard of the possibilities of making fabulous hauls of diamonds from the seabed off the gaunt coastline of SWA, he did not rush in where the average prospector would fear to tread.

"He proceeded with caution, and made a thorough investigation of the prospects of striking it rich beneath the restless waters of the Atlantic Ocean. He was soon convinced."

South Africa rife with rumours

While this was all quite true, it took some time for the full story of the real-life drama being enacted out there off the desert coast of South West Africa to get through to a largely sceptical public ashore.

Before the rich hauls of sea diamonds by *Barge 77* were disclosed, and verified, South Africa was rife with rumours that the entire sea diamonds project was a gigantic swindle.

Piet Beukes, who as a journalist and editor remained close to Collins and the sea-diamonds story throughout, recalls that "everyone had expected the Texan to announce big diamond discoveries, and when none came to light, and Collins could not be found in his office, the rumours spread.

"At the same time there were those who predicted that Collins would eventually turn up with diamonds, but that these would have been planted in a certain place where he would later 'discover' them."

In May 1962, about the time that *Barge 77*, on its way up the West Coast, was towed by the *Emerson K* to a site near the Olifants River mouth for preliminary tests, Collins had bought the controlling interest in a small West Coast diamond mine in that area owned by a company registered as Weskus Mynbou.

"Weskus Mynbou," says Beukes, "had been the source of much speculation and rumour.

"In 1946 a South West farmer, Lukas Steenkamp, who had previously won and lost a fortune, had discovered diamonds by accident at the mouth of the Olifants River, when he and his brother were on a fishing trip in the area.

"Later, Lukas and his son Jannie formed a company to mine titanium, which had also been found in the area. But they soon ran into severe funding problems and in order to obtain finance they sold and gave away shares to a number of people – among them Fasie Malherbe and Jan Haak, an attorney from Bellville who later became Minister of Mines.

"While the Steenkamps and their partners were mining titanium they found a rich pothole full of diamonds, under the titanium. It was subsequently reported that diamonds to the value of R1,4 million had been taken out from this area.

(It was in the sea-diamond concession north of the Olifants River that a 36-year-old marine geologist of the Anglo American Corporation, Dr Joseph Andrew Wright, later disappeared while doing underwater geological tests. His body was not recovered).

Oppenheimer's offer

"The quarter-of-a-million shares in Weskus Mynbou suddenly became worth a lot, and it was reported that Harry Oppenheimer had offered one million rands for a majority holding.

"At this stage, Sam Collins appeared on the scene with an offer the equivalent of R1,2 million for 14 000 of the shares. The Steenkamps sold their shares, and various other people made small fortunes out of the deal.

"Kappie Strydom, who was chairman of Weskus Mynbou at the time, was in London and he was not aware of this deal until notified by cable. This led to big complications later on.

"Collins, with typical haste and hurry, immediately set about developing Weskus Mynbou as a first-class mine. He invested R300 000 to acquire new machinery and to increase production.

"The fact that Collins had taken over this diamond mine along the coast immediately gave rise to rumours that he was anxious to have a

The motor yacht Schipa, *used initially as a crew-change ferry, at anchor off the SWA coast.* Barge 77 *in background.*

source of diamonds so as to be able to plant them in the sea, where he could 'discover' them with his pilot barge. Cape Town seethed with rumours, and the most extraordinary things happened behind the scenes.

"Lukas Steenkamp had earlier discovered a second diamond area at De Punt, in the vicinity of the Weskus mine, but as he was short of money he sold his interest in De Punt for a mere R12 000. Shortly after the Weskus deal with Collins, De Punt was bought by a Dutch/Belgian group for R1,2 million. I picked up the news and heard that Jan Haak had a large interest in De Punt, which he sold.

"Three weeks after my story on De Punt, I was able to disclose that a former National Party MP, Daantjie Scholtz of Springbok, had acquired a diamond concession on a farm at Wolfberg, on the Elands River in Namaqualand. He was reported to have taken out diamonds to the value of more than R1 million before he sold Wolfberg to the same Dutch/Belgian group that bought De Punt.

"These and other developments created intense excitement, and the rumours became wilder and more widespread by the day. All kinds of strange people with diamond propositions and 'inside' information turned up at my office. My phone, both at the office and at home, rang incessantly, with inquiries about what was going on.

"I phoned Sam Collins's office practically every day to try and get some reliable information on *Barge 77*. But Collins at this time was away, and every time I got the same reply: 'Mr Collins is not available'.

"A high government official then contacted me from Pretoria, to ask if I could help with information on the sea diamond ventures. He said the government was very worried, as the rumours that were circulating would only harm South Africa's reputation in financial circles abroad.

"I subsequently heard that the Prime Minister, Dr Hendrik Verwoerd, had taken a personal interest in the matter, and was asking for information. Dr Nico Diederichs, then Minister of Economic Affairs (later Minister of Finance and finally State President) was in London at the time, and we heard that he too had been making inquiries.

"I then learnt that arrangements had been made for certain financial groups to buy back Weskus Mynbou from Collins, and that this had the approval of government. It became obvious that there were manoeuvres and counter-manoeuvres behind the scenes, to prevent Collins from exploiting Weskus Mynbou.

"I heard later that the deal had gone through and that Weskus Mynbou had been drawn into a larger group called Duineveld Beleggings.

"Still no sign of Sammy Collins, and all attempts to get hold of him at his office proved futile. There were even rumours that he had skipped the country – with some of the assets of the companies under his control.

"The MDC shareholders were becoming restless and I heard that, at a board meeting, some of the bigger shareholders had issued an ultimatum to Collins, to produce diamonds from the sea before the end of August – or else!

'Gigi' gives the game away

"It was then that my newsman's instinct and intuition prompted me to contact Sammy's wife, in Sea Point. Mrs Georgette 'Gigi' Collins is a very lovely and charming lady whom Sammy had married in 1956.

"Where Sammy was restless and forever on the move, Gigi as I remember her is a true home-maker, well-known for her culinary magic. A former model, she is also a very talented lady, with a keen interest in the people and customs of foreign lands.

"When I phoned her on that memorable day – it was towards the end of August, 1962 – I reminded her that we had met through the Tretchikoffs. I pointed out that my wife and I lived near to the Tretchikoffs, and invited Sammy and herself around for a drink.

"Sammy, she replied, was away.

'Do you ever hear from him?'
'Oh yes, he talks to me on the radio every day, from the barge.'
'And how is he getting on?'
'Fine.'
'And have they discovered any diamonds?'

'Oh yes, plenty of them. Sammy told me they had a wonderful haul of diamonds from the sea at Chameis Bay. He told me the whole undertaking had been most successful – more so than they had dared hope for.'"

Gigi declined Piet Beukes's invitation with regret, but said that when Sammy returned, they would be pleased to visit Piet and his wife at their home in Constantia.

"My talk on the telephone had served its purpose," says Beukes.

"I had the story I had been looking for and on August 25, 1962, six days before the shareholders' ultimatum to Collins was due to expire, I was in a position to announce to the world that Sammy had found a rich haul of diamonds in the sea, at Chameis Bay.

"My newspaper (*Die Landstem*) appeared with the news – under my personal by-line – on the Wednesday morning, and that afternoon the shares of both General Mining and West Wits, the company through which Anglovaal had taken an interest in sea diamonds, showed an appreciable increase.

"Sammy meanwhile had left *Barge 77* and was returning to Cape Town in his crew-change ferry, the *Schipa*, with the first diamonds sucked up from the sea from the barge, in a little plastic container that he kept in his pocket.

"There was a storm on the way, and *Schipa* was delayed nearly three days. When Sammy eventually arrived in Cape Town he learnt that I had scooped him on the big story he had hoped to announce himself. When he found I had got the facts from his wife, he did not take her to task.

"Instead he regarded it as a big joke, that his wife had disclosed the big news before he could, and that I had tricked her into revealing what he had told her in such great confidence, over the closed circuit of their radio link.

"When I saw Sammy later he said with a smile: 'Piet, you sure stuck your cotton-pickin' neck out with that report – just as I stuck mine out by lookin' for diamonds in the sea!'"

A "glittering new sea harvest" off SWA

The Cape Editor of the *Financial Mail*, Anthony Heard (later to become editor of the *Cape Times*), reported that "a glittering new sea harvest – diamonds – is being reaped off the South West African coast. The companies pioneering this project are already seeing results which promise to cover current prospecting and mining costs."

The *FM* quoted Sam Collins: "We have verified the existence of diamonds in numerous places along the coast from the Olifants River to Luderitz. But the yield we can expect is not yet proven."

But, the journal added, Collins and his colleagues were "extremely happy" about the initial yield of a particularly fine quality of diamond. "They are hoping for a 'strike' by finding a diamond-rich 'pipe' like those found on land."

The *FM* said total investment so far in the project had been nearly R2,9 million, and that the group hoped to bring additional mining units into operation – and that until this was done "it is doubtful that it will be able to declare profits."

In March 1963 the journal reported that in just over a fortnight's hectic trading Sea Diamonds – not yet quoted – had moved from 70c to 180c. "Now that Mr Collins has his money securely in the bank he will probably clear the way for a listing. That would mean a further issue to bring the public's stake up to the minimum required of 20 percent of the share capital.

"Here it would be as well to remember that dividends from Sea Diamonds will probably not start flowing for a couple of years. A lot of money will be needed for equipment; the big mining rig Mr Collins wants to instal will cost about R1,5 million. But he is confident that with it, he can produce between 1 000 and 2 000 carats a day.

"Since he started operating from his small barge, profits have totalled R160 000. Last month's profit was about R60 000 which, with present equipment, is expected to improve slightly.

"Stones are averaging 55 points compared with an average 86 points at Consolidated Diamond Mines. The largest stone fished out so far has been 11 carats – but not a good stone. The most attractive has been an 8,5 carat blue-white.

"The bulk are of good quality."

De Beers's R2 million loan to Collins

Marine Diamonds' production attracted the interest of De Beers, and early in 1963 Harry Oppenheimer announced that the group had entered into an agreement with Sam Collins in terms of which De Beers had undertaken to lend R2 million, partly to Collins personally and partly to the Sea Diamond Corporation which he controlled.

It was also reported at the time that General Mining and Anglovaal would likewise put an extra R1 million into the sea diamonds venture.

In announcing the De Beers loan, Oppenheimer said its purpose was "to assist in the financing of two companies, Marine Diamonds Limited and Southern Diamonds Limited, which hold certain rights to recover diamonds from the sea off the South West African and Namaqualand coasts.

"Diamonds have been found in prospecting at many points in this area, and Marine Diamonds Ltd is at present producing on a considerable scale from the sea off Chameis Bay in SWA and is planning to increase the scale of operations substantially.

"No production has yet come from the area along the Namaqualand coast. It is too early at this stage to assess the importance of these projects with any confidence, but it is quite possible that either or both of them may in time become important sources of diamonds.

"Certainly, Mr Collins's work in recovering diamonds in these novel conditions is a technical achievement of a high order.

"The loan agreement we have entered into with Mr Collins provides, inter alia, for an option to us in certain circumstances to acquire part of Mr Collins's interests (which amount in total to about 40 percent of the capital of each of Marine Diamonds Ltd and Southern Diamonds Ltd) and a right of first refusal, subject to certain prior commitments entered into by Mr Collins, on any further interest in these companies of which he may wish to dispose.

"Our agreement also provides that Mr Collins will use his best endeavours to ensure that any diamonds that may be produced by Marine Diamonds Ltd or Southern Diamonds Ltd will be marketed through our sales organisation".

The influential *Mining Journal*, London, asked editorially at the time:

"What does all this mean? De Beers appears to be unwilling to amplify its announcement. But according to the London end of the Collins organisation, the £1 million (R2 m) is in effect being lent to Sea Diamonds which owns the current Collins stake of 43,75 percent in Marine Diamonds. General Mining and Anglo-Transvaal jointly hold another 43,75 percent. The other 12,5 percent is understood to be held privately by interests at the Cape.

"What is not clear is why Mr Harry Oppenheimer and De Beers have decided to go into this venture now, when they have hitherto been holding resolutely aloof, and what strings are attached to the loan. One thing seems to be fairly certain: the Collins diamonds will be marketed through the De Beers Central Selling Organisation.

Miners were transferred from ferry to barge in this tiny dinghy. Many were so petrified they refused to board it.

"So far only one experimental barge is at work recovering the diamonds. It began serious operations last August and is now believed to have worked up to an output which on some days has been passing the 700 carat mark.

"This is, of course, small compared with the huge production that De Beers Consolidated Diamond Mines of South-West Africa obtains from the coastal sands in this region. But Collins has so far proved two things: that the diamonds are there, and that they can be sucked up commercially.

"The next move is to put other and larger units to work. One such is planned to start up later this year. Some £2 million has already been sunk into the venture.

"The only direct public participation so far has been through the one-shilling shares of Sea Diamonds, which have been made available in Cape Town at prices decided by negotiation, there being no stock exchange quotation for them. The price in December was 7s 6d. Now it is nearer 18s.

"With three major South African mining houses now in the swim, what was formerly regarded as something of a swashbuckling adventure is becoming a serious mining operation indeed."

It so happened that, just as Sam Collins was beginning to demonstrate that technically the winning of diamonds from the seabed was a practical proposition, a Bill was introduced in the South African Parliament, in April 1963, extending the territorial waters of South Africa and SWA from three to six nautical miles.

Although this new legislation had been planned long before a start had been made in the sea diamonds venture, it fortuitously had the effect of doubling the size of the offshore zone over which the SA government had jurisdiction.

The Durban City Council pipeline contract

Meanwhile Collins Submarine Pipelines Africa (Pty) Ltd had been awarded a contract by the Durban Municipality to lay an underwater sewage outfall pipeline at the Umlaas Canal on the Bluff, near Durban's international airport.

Captain George Foulis recalls that not long after he had been put in command of the *Emerson K* he was sent to Durban to look at how the survey was being done for the pulling of this pipeline.

"Several of Collins's American associates in his pipeline business were in Durban at the time in connection with this project – people such as Billy Glasscock, Wynn Harvey and Bill Evans.

"I remember being told at the time that Sammy Collins had succeeded in acquiring a very unusual type of insurance policy; one that was difficult to get at the time.

"This policy apparently had the effect of insuring him against bankruptcy in that, in certain specified circumstances, he would be able to fall back on it to enable him to remain solvent.

"As it happened, heavy storms struck the Natal coast during the course of the pipeline contract, and these washed away part of Collins's launching jetty and part of the pipeline. Sammy was running so short of money at the time that he was unable to pay his workers.

"It became clear that these employees would not continue working on the contract much longer if they were not going to be paid. The terms of the special insurance policy apparently did not help him in this respect, so Sammy found himself in a difficult situation.

"The Durban City Council was not happy with his methods, and felt he was not going about things the right way. After the storm damage the Council said in effect to Sammy, 'this proves you are not competent to do the job, so we're taking you off it'.

"Sammy protested, but he nevertheless moved off the job and was, in fact, only too pleased to do so as he was then able to concentrate on his Marine Diamonds venture. The pipeline contract was given to a major South African civil engineering company.

"Collins subsequently sued Durban Municipality for wrongful dismissal and for breach of contract. He was able to prove in court that, far from being incompetent in this field, he had successfully pulled bigger and longer submarine pipelines in other parts of the world including America, India, the Mexican and Persian gulfs and Gibraltar.

"On the strength of proven facts that he was able to produce in evidence, the court ruled that Collins could not be found to be incompetent or inexperienced in the construction and pulling of submarine pipelines, and it found in his favour.

"The outcome was that Sammy received an extremely good damages award – presumably for wrongful dismissal from the contract."

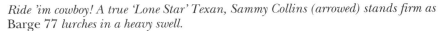

Ride 'im cowboy! A true 'Lone Star' Texan, Sammy Collins (arrowed) stands firm as Barge 77 *lurches in a heavy swell.*

On one of the calmer days off the West Coast, Collins, a highly-experienced diver, goes down to look for himself.

CHAPTER SIX

Little-known facets of a rough diamond

SAMMY COLLINS was a dynamic, decisive and hard-driving bundle of energy, and Socrates Vartsos says that working with him was "like living alongside a volcano".

Vartsos, as a mining geologist, not only worked closely with Collins; he also got to know him well socially.

He describes his former boss as "a real hands-on manager; a man who worked as hard as he expected those around him to work, and who would take instant decisions – and stand by them.

"He was like a multi-faceted diamond; a man with a many-sided make-up and with a spirit and personality much bigger than his body. He was the sort of person who, once you had had close contact with him, and worked with him, you would remember for the rest of your life – unlike the 'grey' people one so often meets these days.

'He was a self-made man who in the end also self-destructed. In his lifetime he must have swung from bankruptcy to being a millionaire at least three times. He would lose everything and in no time be right back on top of the heap again. His disrespect for money gave him an ability to take chances with it, where others would cautiously hang on to their cash as a form of security."

Sammy Collins was once also described, by a newspaper interviewer, as "a man who builds fortunes with the same grave relish of a small boy mixing mud-pies".

He was a dollar multi-millionaire before he came to SA to lay undersea pipelines and hunt for sea diamonds, so he could have retired in heady luxury before his African venture. But, for a man who liked working hard and living hard, such a life would have been sheer hell.

Collins made it clear, in an interview in the early 1960s, that it was not a love of money that kept him making more and more of the stuff. It was simply that he loved working – and the harder and more challenging the work was, the more he liked it.

To this man who at one time was reckoned to be worth at least R10 million in personal holdings in his companies, big business was a game, and money was the stuff one played it with.

He told his interviewer, Brian Barrow of the *Cape Times*:

"Okay, you can be lousy with money. You can buy whatever you want, but when it comes to the pleasures of life you can only drive one car at a time, wear one suit at a time, eat one fat steak at a time and drink one scotch at a time.

"And if the best things in life are free, you can't enjoy them more than anyone else."

During his brief sojourn at the Cape, Sammy certainly left no doubt that he liked the good things of life. After a hard day's work, and keeping in constant touch with his mining and other interests, he liked nothing better than to relax with friends in an up-market night spot.

Maxime's on the Cape Town Foreshore was a Collins favourite, as was the Sable Room with its panoramic view of Table Bay from the top of what was then the tallest building in Cape Town, the Sanlam Centre on the Foreshore – later to be renamed Nasionale Pers Centre.

"A man's man – and a ladies' man..."

Sammy has been described by a close associate as "a man's man – and a ladies' man", who liked to be in the company of pretty women.

His wife Gigi once laughingly remarked to an interviewer: "When Sammy was in oil, nobody looked at him. Now he's in diamonds, all the girls suddenly love him!"

He was also a keen follower of horse-racing, and a frequent punter at racecourses in the Cape Peninsula.

Collins was married three times and he had three children, a son, Sammy Junior (who did not follow his father into the worlds of oil and diamonds) and two daughters, one of whom, Stephanie, became a Catholic nun in Los Angeles after the break-up of her parents' marriage.

In Cape Town, Sammy and Gigi lived in a company apartment in Chartleigh House on the seafront at Three Anchor Bay. He was driven around in the company's black Cadillac, by a chauffeur called Louis, and his hefty personal bodyguard, "Fido" Brouwer, was always near at hand.

The black Cadillac is the subject of some of the countless anecdotes concerning Sammy that still do the rounds among those who worked for him or who knew him socially. One of the stories told is that Collins boasted, Texan-style, that his Cadillac was faster than his co-director Bill Mitchell's Rolls-Royce.

Mitchell, also a stocky self-made millionaire but as British as Collins was American, took him up on this and in the early hours one morning – after a customary session of night-clubbing – the two tycoons raced their respective limousines along a stretch of open road outside Cape Town.

The Rolls-Royce won, and Sammy was so put out by this that he got rid of the Cadillac and bought a later – and faster – model in the Cadillac range!

Fido the bodyguard, as tall as his boss was short, joined the Collins entourage at an early stage, when the Marine Diamonds offices were in the Radio City building on Cape Town Foreshore's Tulbagh Square. Socrates Vartsos recalls that the daily sight of the squat, balding Sammy, shadowed by his towering bodyguard, never failed to bring smiles to the faces of MDC staff.

While in many respects fitting the soap-opera image of a Texan oil baron, Collins tended to dress conventionally and although he had a small pearl-handled automatic on his leather belt and sometimes wore Western-style boots, he did not sport the "ten-gallon" Stetson that has become the hallmark of tycoons from the Lone Star State.

Nor did he chomp fat cigars. He did, however, smoke cigarettes – and always an American menthol-cooled brand.

His favourite tipple was Johnny Walker Black Label whisky (Red Label too, sometimes) and Riebeeck Water. He would have his first whisky of the day at 10 am when he was at the office, and he would often get through at least a bottle by the end of the day.

An "amazing" capacity for alcohol

"Socs" Vartsos says Sammy amazed everyone with his capacity for alcohol. "He'd arrive at the office at 10 am, and while sipping his first whisky he would get on with the morning's work and take what-

From the control cabin on Barge 77, *geologist Socrates Vartsos makes a position report to the shore-station.*

ever decisions had to be taken. By lunch-time the level of the whisky bottle would already be well down.

"He would then go across to Maxime's for lunch with business and other guests, and he'd have some wine with the meal. Then, having said goodbye to his guests, he would go across to his town flat in the Radio City building and he would have a zizz for about two hours.

"Then, about 4 pm, he would bounce back into the office just when everyone else was winding down towards the end of the working day, and he would now want to get to work, and to confer with his staff about the day's production, operational problems and so on.

"We rarely got away before 6 pm. Sammy would then start an evening of socialising. He would have pre-dinner drinks with friends – quite often including myself – followed by dinner at the Sable Room or some other night spot. This happened quite regularly, and Sammy would rarely go home before midnight, or early next morning. And we would all have to be at the office by 8 o'clock!

"This was almost a daily routine. Sammy really lit the candle at both ends. He had boundless energy and an impish sense of humour, and he was fun to work with. But in spite of the first-name relationship we enjoyed with him, no-one took advantage of it – and no-one ever doubted that he was always the boss.

"He could really tear strips off people for not doing their jobs the way he wanted them done. I remember for instance Sammy being driven to my flat in his Cadillac, and Louis his driver doing something quite uncharacteristic in trying to turn this big car round in my narrow driveway.

"The car became completely stuck, between a rock-face and the building, and we ended up having to get a tow-truck to pull it out of the driveway.

"Sammy was of course furious. He shouted all sorts of obscenities at poor Louis – and he fired him, there and then. But when Louis called at the office next morning to collect his final cheque, Sammy merely said 'get the car and let's go!'

"This was how Sammy was. He never harboured grudges. When he did something he may have regretted afterwards, he would be apologetic the following day, or make amends in some other way – or just brush an incident aside as if it was not something to be offended or upset about."

Socs Vartsos remembers that there was also a dark side to Sammy Collins.

"He could be uncontrollably jealous in his relations with people – particularly with the ladies. And he didn't take kindly to being questioned about decisions he had taken – even if one knew that what he was doing was not right.

"In that sense he was stubborn and obstinate. And these are the

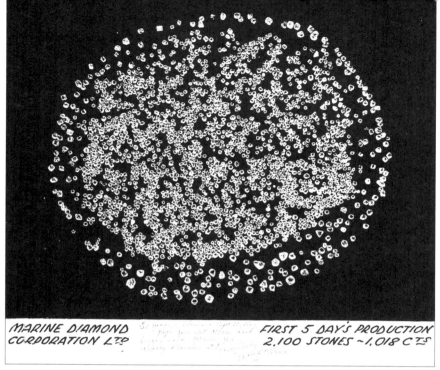

This mounted photo, signed by Sammy and sent to Johann Vivier in September 1962, shows Barge 77's first big haul. Worth less than R30 000 at the time, a parcel of diamonds this size would fetch close on R1-million in the 1990s.

traits that brought Sammy into direct conflict with management of the Anglo American Corporation – because those managers were trained to be anything but stubborn and obstinate. They were meant to have flexible minds; a pragmatic approach.

"Certainly, there was a side of Sammy Collins that you didn't want to know – especially the side that came out after midnight. But then don't we all have our darker side?"

Another of Collins's close associates, Tom Kilgour, goes so far as to refer to his former boss as "very uncouth, with crude mannerisms such as the way he cleared his throat, and spat to one side". Yet his overriding recollection of Sammy is of "a human dynamo who was also a very sincere and generous person, who would do anything for you if he liked and respected you.

"One either loved the man or you hated him; and if he took a disliking to you, the best thing to do was to keep well out of his way!"

"Sammy woke up a sleepy Cape Town..."

With the expansion of Marine Diamond Corporation's operations, Sammy moved his headquarters from the Radio City block to the neighbouring Barclays Bank building on the Foreshore, facing onto the Heerengracht.

Sidney Kagan, a Cape Town accountant, was among those who

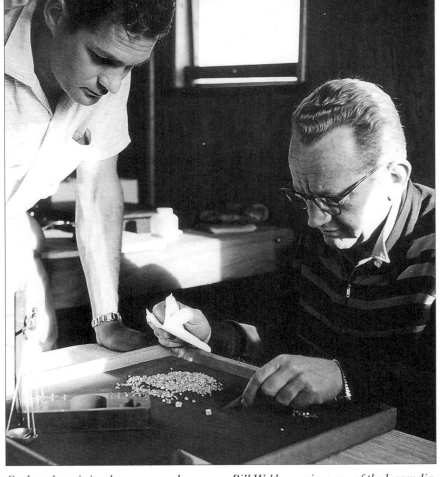

On board a mining barge, general manager Bill Webb examines one of the larger diamonds recovered from the seabed.

joined the fledgling MDC in its Radio City days. Just before his sudden death at the end of 1995 he recalled in an interview how "very exciting" it was to be working for the dynamic and unorthodox Sammy Collins after the "humdrum" routines and constraints of accountancy.

"It was an entirely different world from the one we were accustomed to; none of us 'locals' had ever experienced anything quite like this before – and at first it came as a shock to the system.

"Sammy obviously thought of us as yokels living in the backwoods, and that he could take us all on – and he did. He had his own way of doing things, and he went right ahead and did things that way; it was no good trying to tell him otherwise.

"And there can be no doubt about it: Sammy Collins woke up a sleepy Cape Town. He really got things moving here, and a lot of individuals and a lot of companies did very well out of him.

"Sammy had no idea at all of accounting procedures. There were no such things as motivating talks, or anything like that with him; if there was work to be done we just had to get on and do it – and that was that.

"Whenever I think back on those days I'm reminded of just how incredible the Collins era was. Money meant nothing to Sammy, and we had a really torrid time trying to balance the books and to take stock – and account for stock.

"Motor vehicles would just disappear overnight, and no-one could tell us what had happened to them. In the end we just had to write them off. And when the barges were being built, welding units would be simply tossed over the side, and we would have to dredge them out of the harbour.

"Before the first barge, *CCC 77*, was converted for diamond mining it was estimated that the conversion would cost one million rands. This soon turned out to be a miscalculation, and the ultimate cost was nearer two million rands. But no-one worried about this, least of all Sammy; to him this was just a small error, and nothing to be concerned about.

"That was the way things happened – and they happened fast! I pitied the poor auditors. What a job they had!

"At first, when the diamonds arrived at our offices from the barges, and after they had been recorded in the books, I would take them across Adderley Street to the Standard Bank, in a small case. I did this regularly, and I had no armed escort and no special security precautions were taken. In those days, you could do these things!"

Sidney Kagan recalled that one day after Marine Diamonds had moved from Radio City into more luxurious and spacious accommodation in the Barclays Bank building, he heard the sound of shots being fired down the passage.

"I hurried out of my office to see what was happening, and there, in another office, I saw Sammy Collins, his son Sammy Junior and the artist Tretchikoff taking pot-shots with a pistol at a smoker's pipe which they had propped up against the wood-panelled wall. There were bullet-holes all over the beautiful woodwork!

"It turned out that the trio were putting in some practice for a hunt they were all going on soon. And this practice shoot was before 10 am, when Sammy used to have his first whisky of the day!"

Kagan, who worked under Collins's financial director, Danny Ipp, said that periodically he would have to go to SWA to be flown out to the barges to take stock. "For the few days we were there we lived like kings. The food on the barges was unbelievable – real five-star quality.

"But as far as the stock-taking side of it was concerned, this was something of a nightmare. Nothing ever agreed or balanced, and we had the devil of a job trying to locate equipment that was supposed to be in stock."

The Clifton Hotel pistol-pulling incident

Sam Collins's flamboyant lifestyle frequently got him into the newspaper headlines – and sometimes into trouble.

George Foulis was with him when Sammy pulled his pistol at the Clifton Hotel – and had a brush with the law.

"Sammy had invited some of us for drinks at the hotel one late-afternoon. We took a table, and kept a few chairs for more of our people who were expected to join us later. There were some other guys sitting at a table nearby, and one of them came over and asked if he could take these empty chairs from our table, for people *they* were expecting.

"Sammy said a very emphatic 'no', and explained why we needed these extra chairs. About half an hour later, when the people we were expecting had not yet turned up, these guys at the nearby table again came over and asked if they could take our unused chairs.

"When Sammy again refused to allow this, these guys said 'to hell with it, we're taking the chairs anyway!'

"At this, Sammy pulled his pistol – the same one he had used to quell the mutiny at Luderitz – and pointed it at these people who were taking our chairs.

"Someone in their party called the police. They wanted Collins arrested and charged with pointing a firearm. But somehow Sammy, through his connections, had the whole thing dropped, and that was that.

"Life was never dull with our Sammy around!"

He called his financial empire "a load of junk"

Sam Collins's visitors, after he had moved to the Barclays Bank building, would find him sitting behind the long curved desk in his 12th-floor office from which he ruled a multi-million-rand financial empire that spanned the world and which he referred to as his "load of junk".

The visitor soon became aware of the chattering sound of telex machines and of the voice of a receptionist-telephonist handling calls to and from New York, London, Paris, Dubai, Toronto, Mexico City, Johannesburg and other places in which Sammy had business and financial interests. At one time he even had an office in the Magellan Strait, near the notorious Cape Horn!

One also became aware of the huge diamond (said to have been around 5 carats, cut) set in a ring that the Texan wore on a finger. This was said to have been the biggest gem found at the outset of his operations off the West Coast.

Reference to this find was made in the September 1963 issue of the *Diamond News and SA Jeweller.*

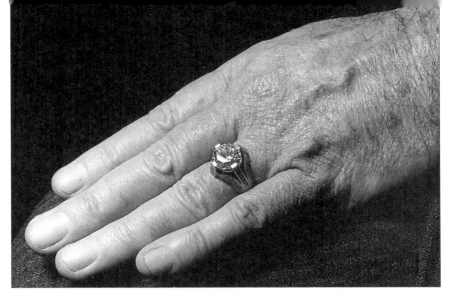

Sammy Collins's diamond ring. The stone, of about 5 carats, was found early in the search off the SWA coast.

"In salvaging the plant from *Barge 77* which was wrecked at Chameis Bay north of the Orange River mouth on July 1," the journal reported, "Mr Collins's men found in the barge's ball mill the biggest diamond recovered since they began prospecting on the West Coast about two years ago.

"It must have been picked up by the barge less than three hours before it was beached, Mr Collins said. It was a blue-white diamond worth a lot of money. But Mr Collins does not intend putting it on the market. 'I am keeping it as a souvenir,' said the 50-year-old millionaire."

(A full account of the grounding of *Barge 77* appears in a later chapter).

Another eye-catching feature of Sam Collins's attire was the smaller diamond that twinkled from his gold tiepin. This was shaped like a deep-sea diver's helmet – the distinctive symbol, or logo, of Collins Submarine Pipelines and also of the Marine Diamond Corporation.

The diamond on the tiepin was inset where the glass face-plate on the diver's helmet would be. Collins gave similar tiepins to his closest associates, with matching cufflinks.

One wonders how far a latter-day Sammy Collins would get, carrying the wealth that he did on his person 30 years ago, in trying to walk down Cape Town's Adderley Street without being mugged. Apart from his diamond ring, the buckle of his leather belt was studded with diamonds, and he also wore a diamond pendant under his shirt. And he got away with it!

At the office, if Sammy was in his usual expansive mood, he would take a tray out of his office safe and show his visitors a collection of the sparkling "ice" that his barges were extracting from the sea. In those days, in the mid-1960s, a display of 1000 carats of diamonds would have been worth a mere R27 000; 30 years later, with the market price at around 200 US dollars a carat, the value of that same display would be close to R1-million.

"I'd go cotton-pickin' crazy...!"

When Sammy spoke, it was with a deceptively slow, couldn't-careless Texan drawl, which tended to mask a man as tough and high-powered as they came.

"I'd go cotton-pickin' crazy if I wasn't on the move all the time, doing somethin'," he once said. "I get a kind of a kick out of it!"

The term "cotton-pickin'" was very much part of the Collins lexicon, and reflected Sammy's Southern background.

He confessed that he found what he was doing to be great fun. "I get lots of enjoyment out of setting myself objectives, gaining them and then going on to new objectives. It's not only my work; it's also my hobby and my relaxation. I have no outside interests and I don't care for any kind of recreation. Ask me to get on a golf course and knock a little white ball around and you'd drive me crazy, real cotton-pickin' crazy."

Of his offshore diamond mining venture in SWA, a confident Collins said: "I find the challenge of underwater projects irresistible. I believe this will be the biggest diamond discovery ever made. I think it will provide gemstones for all the foreseeable future.

"On yield per ton of dirt, it's the richest diamond mine in the world. I just cannot estimate the total yield in our area."

Short, balding and wearing glasses which he would frequently prod back to the bridge of his nose, Sam Collins left no-one in doubt that here was a man who knew what he wanted – and was going for it. And while to the general public he may have come across as a rough diamond; as a boozer, a womaniser and a compulsive gambler, he commanded tremendous loyalty and respect among those who worked with and for him.

Without exception, former Marine Diamond Corporation people who worked with and under the rough-and-tough little Texan spoke highly of him, in interviews for this book.

Bill Webb, who became one of his closest associates, first as operations manager and later general manager of the MDC, refers to Sammy as "the finest man I've ever met".

Webb says although the age-difference between them was only about six years, he looked upon Collins "as a father".

"And for his part, Sammy regarded the company, and all who worked in it, as his family.

"He expected everyone to work at least as hard as he did, and I recall that he had two mottoes: 'If you can't do it you can't stay!', and 'what you intended doing tomorrow you should have done yesterday!'"

Collins's qualities – some of them little-known outside the workplace – that appealed to his staff were the loyalty that he showed them, his ever-caring attitude towards his staff, his generosity and the fact that he treated everyone the same way, regardless of their position in the company.

Workers quickly warmed to the first-name relationships they enjoyed with their boss – but this in no way lessened their respect for him.

All spoke of Collins as a hard driver, but as a fair-minded person who went out of his way to ensure the welfare of those who were pre-

The view the offshore miners had of their environment – the bleak, inhospitable and rock-strewn shoreline of SWA.

pared to work hard and remain loyal to him, and to the company. And he would never assign to any of his staff a task he would not be prepared to do himself. In this way he gained tremendous support, by example.

Captain Hugh Pharoah, a Marine Diamond Corporation pilot in the 1960s and, at the time of writing, a Boeing captain with Singapore Airlines, describes Sam Collins as "a man of startling contrasts".

"He could be very hard, but he could also be remarkably generous on occasions. For example, when I came to leave Marine Diamonds I was flying the company's Twin Bonanza for Sammy's subsidiary company, Argus Oil Exploration, based at Lake Sibaya in north-eastern Zululand.

"When South African Airways contacted Sammy in Cape Town and told him I was required in Johannesburg urgently for a selection board and interview, Sammy pulled all the stops out to ensure that I got to Jo'burg in time.

"He got a message through on the radio link, instructing me to fly the company's aircraft to Jo'burg, do the necessary with SAA and then bring the aircraft back to Cape Town for a crew-change.

"Sammy not only paid all my expenses for the whole trip, accommodation included; he then released me from Marine Diamonds immediately, with a bonus of two months salary and the promise that if things did not work out for me with SAA, I could come straight back to him!"

Collins's Marine Diamonds staff was drawn from a wide spectrum of society. There were township "toughies" who were recruited as labourers or deckhands; there were the office staff and accountants; there were the certificated mariners, miners and engineers; there were the aviators and there were the security staff and the scientific specialists in various fields.

While most of them were recruited locally, at the Cape, some came from overseas to take part in the venture. These included a few fellow Texans who had worked with Sam Collins in his submarine pipeline company – men such as Wynn Harvey, Leroy Rodin, George Guthrie and Bill Evans.

With Sammy, they all contributed to the awakening of a "sleepy" Cape Town.

CHAPTER SEVEN

"A great leader – and a great boss..."

"SAMMY COLLINS", says Francois Hoffman, "was a great leader and a great boss; one who inspired people.

"But I doubt if his style would work in today's world.

"He was a hard driver, but he looked after his staff. He never tried saving money on food, or on crew-comfort. To him, there was no such thing as a vacation, or leave. 'What the hell's a vacation?' he would say. 'There's no such thing as a vacation in my vocabulary! You can take leave when we have nothing to do. But while we've got something to do, you've got to work – and work like hell!'"

Hoffman says that in his first year with Marine Diamonds – in 1961/62 – he slept at home for a total of only 40 nights. "We were forever at sea. This happened to all of us, yet somehow we all came away from it thinking that our boss was a great guy.

"He set the pace himself. He was at the office every day including Sundays. On a Sunday we were expected to be on the job from 10 am to noon, at least. We would talk to the boats to find out what was happening and what their needs were – and the chaps on the boats knew we would be at the office.

"Although he was a hard driver, and although he would chew your head off if he thought you were not pulling your weight, he could also be very nice. Out of the blue, he would say 'Hoffie, you've had a tough time; give yourself a break'. And he'd give me air-tickets so that I could take my wife to Durban. That sort of thing.

"There are a lot of little-known things about Sammy Collins, such as that it was his idea, during the sanctions period, to store oil in stoped-out areas of disused coal mines in the Oogies area near Witbank, in the Eastern Transvaal. I was involved in that too.

"He convinced the government that this was the thing to do, where oil reserves were concerned. Somehow, he did not personally benefit much financially from this. The scheme was eventually taken over by the Strategic Fuel Fund.

"Sammy also had an engineering company in an industrial area of the Cape Peninsula, to service the needs of Marine Diamonds and the Collins Submarine Pipeline Company. In one way or another, a lot of people benefited from Sammy's ventures, and he provided work for many, direct or indirectly.

"He was an easy touch for people with ideas, or with a mining title, and he was forever the innovator, and the entrepreneur. One of his failings was that he drank too much. This became more and more of a problem in the end, and it became known that if you didn't see Sammy before 10 in the morning, it was hopeless."

"No respect whatsoever for money..."

Dr Piet Neethling has described Sam Collins as "a go-getter, if ever there was one. Give him a job to do, or an idea to pursue, and he would go for it, with all he had.

"But he had no respect whatsoever for money. Money to him was just a commodity; a means to an end but not an end in itself. It didn't matter to him if something cost R100 000 or R100 million. If in his view it was worth pursuing, he'd go for it, regardless of the cost.

"Sometime in the period 1962/63, when things were going extremely well for us, Sammy decided to throw a party to celebrate our success – and when Sammy gave a party it was always the biggest and the best – real Texan style. He never bothered to count the cost.

"He invited the whole darned world to this party. Just about everyone in Cape Town was there. It cost thousands of rands, at a time when just one thousand was considered a lot of money."

(It was about this time, when MDC appeared to be doing extremely well, that Collins in a reported interview said in a reference to the critics and sceptics: "You tell them cotton-pickin' bastards that I got more diamonds than anyone in the whole world!")

Piet Neethling says he, Abe Bloomberg and AP du Preez became so concerned, as shareholders, about Sammy's lack of regard for money that he spent some time working in the MDC offices on the Cape Town Foreshore – "just to keep an eye on our interests".

"We thought he was going a bit overboard, and stretching the money a bit too far – money that wasn't really there.

"We became very close friends, Sammy and I, and we and our wives used to visit one another's homes quite frequently. We all got on very well together, and Sammy would philosophise on life from time to time.

"On one occasion he said: 'Piet, what do you really want out of life?' I said that first I wanted to be happy. Secondly I would like to be in a position to provide properly for my family, and thirdly I'd like to have money put away so that one day my children can inherit something worthwhile.

"Sammy gave me a withering look and said: 'Piet, you're a bigger bloody fool than I thought you were! Financially, you owe your kids nothing. What you do owe them is to set them a good example, and

Peter Keeble (left), journalist Piet Beukes (centre) and Sammy marvel at the gem quality of offshore diamonds.

to give them a good education. Nothing more.

'Your having done those things, it's then up to them to make a go of life. And if, after the upbringing and education you've given them, they can't make a decent living for themselves, they'll never make a go of life, however much you may try and help them financially."

Gary Haselau of Hout Bay, a film-producer and photographer, and a pioneer of underwater photography in Southern African waters, worked closely with Collins for four years in various capacities, mainly as the Marine Diamond Corporation's public relations officer and photographer.

"I enjoyed those four years tremendously," he says. "I travelled a lot, and the work was always interesting and exciting. I never knew where I was going to end up at the end of the day. I used to have a clean shirt and underwear, and my toothbrush and passport in my briefcase whenever I went to work because I knew I could well be on my way to Durban, London or even Texas by that night.

"An impulsive sort of chap..."

"Sam Collins, then in his forties, was like that; a very impulsive sort of chap. On getting to the office he'd say to me: 'Things seem to be happening over there,' or 'looks like things ain't going so well over there. You better get your ass in a plane right away and go find out what's goin' on – and perhaps bring back some pictures.'

"Sammy set great store in good PR. He always had PR people around him, and he used them a lot in his marketing strategy. He brought in a lot of business through contacts made at presentations and cocktail parties, and by using audio-visuals and photographs.

"He was a very pleasant person to work for. He was as strong-minded as he was physically strong. Like any good Texan oilman, he was always ready for a fight, but one could not have wished to work for a better boss.

"His critics – and there were many – did not see the side of Sam Collins that we did. His door was always open to us, he gave us all a pretty fair block of shares in the company and he was a very generous and caring sort of person. And he had a great sense of humour. There was always a lot of light-hearted bantering around the office.

"At some 'do' or other at the MDC head-office, the conversation got around to art, and Sammy proclaimed, for all to hear, that 'any cotton-pickin' bum can paint a picture; what's so goddam special about paintin' a picture?'

"Peter Keeble, who happened to be very artistic and was noted for his illustrations, took Sammy up on this and he and others locked the boss in a room with some paints and said he would not be let out until he had painted a picture.

"Well, Sammy did produce a picture of sorts. This depicted a diver under water, and he was so proud of it that he hung the picture in his office. Not only that; on the spot to which the diver was looking on the seabed, Sammy stuck an uncut diamond.

"This was a feature of Sammy's office for quite a while. The trouble was that every now and then someone would come into the office and pinch the diamond, and Sammy would have to replace it.

"I don't know if it ever got to the notice of the CID that the odd uncut diamond was disappearing from the office of the Marine Diamonds chief!"

Like any good Texan, Sammy Collins could swear like a trooper. In fact, says Haselau, "he could swear for ten minutes non-stop without repeating himself!

"And like any good oil man, he liked a game of poker. In breaks between work, at the office and on board his barges, he would play poker with the boys. He'd try and get their pay-cheques off them and we'd try and relieve him of some of his riches. It was all great fun.

"Sammy was a very restless person. He always wanted to be up and doing things. If a project he was working on began to pall, for lack of action, he'd lose interest in it and move on to something new and challenging. At one time I found myself PRO for eight different companies controlled by Collins.

"One of these was the Telenews in Cape Town, which projected the news headlines of the day in moving lights – something like Times Square in New York – from the top of the Sanlam Centre on the Foreshore. This was one of the 'firsts' that he introduced to the city."

(Another "first" that Sam Collins planned for Cape Town was a heliport on top of the Sanlam Centre – but his application was turned down, for safety reasons).

"He didn't mind getting his hands dirty..."

Captain George Foulis remembers Collins as someone who did not mind getting his hands dirty. "He was always a neat dresser, and

A close-up of the pure, gem quality of diamonds taken from the sea off the West Coast of Southern Africa.

although he often wore an immaculately starched, short-sleeved white shirt monogrammed on the pocket, he would not hesitate to get stuck in with the men.

"Those lily-white shirts didn't stay that way for long! One would see him at times shovelling mud and sand off chutes, and handling wet and dirty ropes and other gear.

"On one occasion Sammy decided to demonstrate a point he had made about moving about quickly in the old-type heavy diving suits. I've had a lot of experience with hard-hat divers, and I've known them generally to move slowly and laboriously over the bottom, weighted down as they are by heavy boots and other gear.

"Then Sammy showed us something that impressed us greatly. He went down, in Table Bay Harbour, in one of these hard-hat suits, and put the release valve on an almost-closed position, giving him extra buoyancy.

"And in spite of his heavy boots, he somehow attained a neutral buoyancy which enabled him to move around extraordinarily quickly on the seabed. We could see from the air-bubbles that he was 'walking' the bottom at a terrific pace.

"One obviously has to be very careful how one goes about attaining neutral buoyancy in this way. If you close the release valve too much, the suit suddenly fills with air and before you know what's happening you're spreadeagled on the surface, in your self-inflated 'balloon'.

"But somehow, Sammy had got the adjustment just right – and it worked very well for him.

"Later diving suits had a release-valve control which you activated with your chin."

Hunting Springbok, in the Karoo

Babs Vivier, who married Johann in 1961, remembers that when she was still a young bride, Sammy Collins used to visit the three adjoining merino sheep-farms that her husband owned in the Beaufort West area.

"He liked to come out to the Karoo occasionally, to give himself a break from the city, and to go hunting in the veld with Johann and others from the district. And he loved it!

"From what I have seen of Texas, that part of the world is very similar to the Karoo, with plenty of clear air and wide open spaces, and so Sammy felt quite at home on the farm.

"I found him a very pleasant person – and as a man who liked his food. He loved juicy steaks and seafood – particularly oysters, with tabasco. But where meat was concerned he would eat only beef or venison. He would not touch lamb, for which the Karoo is famous. The reason for this was that in Texas, as Johann and I found when we went there, the lamb has a funny, unpleasant smell and taste, and people just don't eat it there.

"Sammy said he'd been put off lamb for life, by the poor quality of the stuff in Texas."

Sam Collins's social life was as hectic as his working regimen, and no sooner had he and Gigi set up home at the Cape than they were meeting people across a broad spectrum of society in this part of the world.

Sammy's outgoing, fun-loving and dynamic personality drew people to him, and he made many lasting friendships.

Marilyn

One of the many people he met, admired and helped in their careers was Marilyn Human (later Marilyn Martin), an attractive young part-time fashion model, who had come to Cape Town after matriculating in 1960 at her home-town, Heidelberg in the Cape.

She recalls that she first met Sammy Collins in 1961 at a cocktail party in millionaire John Schlesinger's penthouse on the Cape Town Foreshore.

"There was a very active night life in Cape Town at the time, and Sammy and I were moving in the same social circles, with a number of mutual friendships.

"I was then working for Audrey Weinreich's Vogue Academy as a part-time fashion model. At that time this was not a highly-paid job, and it involved appearing at occasional fashion shows and modelling at tearooms in department stores such as Stuttafords and Garlicks. My full-time job was as secretary to the country editor of *Die Burger*. (The Afrikaans daily newspaper in Cape Town).

"I later worked for Aat Kaptein who had founded a newspaper

Marilyn Human (later Marilyn Martin) was prompted and encouraged by Sam Collins to pursue a career in visual art.

called *Die Banier*, which was aimed specifically at the coloured community, and then after completing a course in fashion-modelling I returned full-time to Audrey Weinreich's academy where I did some teaching and modelling.

"While at school in Heidelberg I had taken piano lessons, and while music was the direction in which I was being pointed, I developed a love of painting and began dabbling in art. When Sammy saw some of my paintings in Cape Town he thought they were rather nice, and he encouraged me to develop my love and feeling for art.

"He was that sort of person; a person with an extraordinary faith in the abilities and potential of other people – the sort of person who believed that one could achieve anything that one set out to achieve.

"He was a charismatic person who inspired others. There was nothing superficial about the encouragement he gave people; it was a genuine belief in their abilities and their possibilities. He did things with great conviction, and he had this desire to see people grow, and to develop beyond themselves, as it were."

Encouraged by Sammy, Marilyn enrolled at the Cape Town Art Centre (formerly the Metropolitan Golf Club clubhouse and later to become Seagulls restaurant) on Green Point Common.

"That was a very important part of my life," she says, "because at the art centre I met people like Kevin Atkinson, who not only taught me to paint but also to look farther afield, and to pursue my studies.

"Later, when Kevin opened his own small studio in Strand Street, I joined a small group of people who painted there, and did printmaking and so on – and I was encouraged in all this by Sammy.

"A catalyst in my life..."

"In this way, he became a catalyst in my life – as he was probably a catalyst in the careers of so many others whose lives he touched.

Built in Cape Town in record time, Barge 111 *arrives in Chameis Bay in 1963 to replace* Barge 77 *(wreck on right).*

Cottonpicker, *the first of three Dakotas acquired by Collins to service his mining rigs, arrives at Luderitz.*

"He gave me an exposure to art which put me in touch with others who helped and encouraged me all along the line. He saw in me a very young person who had come from a small country town and also from a very humble background, as he had done.

"We also shared a belief in creating one's own persona; in being responsible for one's own destiny, as it were.

"One remembers Sammy for his generosity and for his warmth, and for his almost childlike faith and belief in people. His exterior of apparent bravado and hardness, and of running roughshod over people, hid a deep sensitivity and a vulnerability that were rarely in evidence in public. He could feel deeply hurt when anyone in whom he had placed his trust let him down."

In 1969 Marilyn married an artist from Scotland, Norman Martin, who was teaching sculpture at the UCT's Michaelis School of Fine Art in Cape Town, and they had two children, John and Catherine. When Norman took up an appointment in Pietermaritzburg his family moved to that city with him. He subsequently returned to Scotland.

Meanwhile, Marilyn had acquired an honours degree in the history of art and also in the history of architecture through the University of South Africa. For a while she taught the history of art at the University of Durban Westville before moving to Witwatersrand University in 1977, where she finished her master's degree in the history of architecture.

She held an appointment in the department of architecture at Wits before being appointed Director of the SA National Gallery in Cape Town, a post which she took up in 1990.

Marilyn last saw Sammy Collins in the mid-1970s, when he was visiting South Africa from his and Gigi's home overseas. "He called to say hello when my children were still very small, and when we were still living in Natal," she says.

She remembers him as "a person of extreme generosity – a generosity of the spirit.

"For someone who had had very little formal education, Sammy had a vast store of general knowledge, and he read widely, with a particular interest in medicine. One could see him as the epitome of the self-educated person; as a self-made man.

"He had a great willingness to share with other people. He didn't take worldly possessions all that seriously, and he never talked much about personal possessions or of acquiring material things and hanging on to them.

"He made a lot of money in his life, and he also lost a lot. It was part of a cycle of having everything one day and nothing the next.

"What he did talk about was the realisation of his dreams, and of being able to take other people along with him, in the fulfilment of those dreams. With Sammy it was never a question of how much money he could make out of diamonds or anything else. It was the challenge that fascinated him – the challenge of doing something that nobody had succeeded in doing before him.

"And with it all, he had a great sense of fun, of *joie de vivre*. He liked the night life of Cape Town which, in the 1960s before the death of District Six and the advent of television, was bright, cheerful and lively.

"Those were the days of thriving night-spots such as Maxime's and the Sable Room – which Sammy frequented – and of other more bohemian places that didn't much attract him but which we in the art world used to go to such as Daryl's, the Catacombs, and the Navigator's Den.

"There can be no doubt that Sammy Collins was a controversial person; a down-to-earth, hard-working, hard-driving, hard-swearing and hard-drinking man who was perhaps not the best-mannered person around. But on the other hand we had here a remarkable visionary and a man of enormous energy; a person not only with a vision but also one who believed in and inspired those around him.

"We all walk paths in our lives, and our paths may cross, and then separate again. Sometimes one is left with a lot of pain, or with wonderful memories. For me, it was indeed a privilege to have crossed paths with Sammy Collins in my lifetime.

"And whenever I think that life is hard, or when I'm battling, I can hear Sammy reminding me of the ancient proverb: 'I used to complain that I had no shoes – until I saw a man who had no feet'"

A mild stroke while airborne – then back to work

Joe and Andrew Horne, former West Coast crayfish divers who joined Marine Diamonds in 1961 as seamen/divers and worked their way up to supervisory positions, both described Sam Collins as "a terrific man to work for".

Joe commented: "If I had to start my life all over again, Sammy Collins is the one person I'd like to work for again.

"His staff were very important to him, and because of the consideration he showed us, none of us minded working long hours overtime for him, or forgoing the free time that was due to us. And

High and dry! Barge 77, its moorings broken by heavy swells, ends up a twisted wreck on the rocks at Chameis Bay.

although he expected a lot from his employees, he paid us a lot more than the average salaries at that time."

Joe recalled that during the 1960s Sammy had a stroke while returning to South Africa by air from the Persian Gulf, to be in Cape Town in time for a staffer's birthday party he particularly wanted to attend.

"In Johannesburg, he was brought off the plane in a wheelchair and taken by ambulance to hospital. While he was receiving treatment there, he suddenly disappeared.

"He had grown impatient and just couldn't be bothered lying there in bed, so he walked out, caught a plane to Cape Town – and got here in time to attend the party.

"He was not the sort of person to allow a mere stroke to stand in the way of something he wanted to do.

"That was Sammy Collins for you!"

The Marine Diamond Corporation's four-engined Douglas DC4 Skymaster, before one of its crew-change flights.

On board Cottonpicker *en route from Cape Town to SWA: Vladimir Tretchikoff (left), Gary Haselau, Rex Redelinghuys.*

CHAPTER EIGHT

On the rocks – on a desert coast!

IT was the "sixth sense" of a German optician, Hans Abel, that led to the first significant discovery of sea diamonds during the 11 months that *Barge 77* worked the MDC concession area in 1962/63.

With fellow diver Joe Horne, he subsequently became well-known in the industry for his uncanny intuition in the search for diamonds on the seabed, and he worked his way up from seaman/diver to mining superintendent, first with Marine Diamonds and then with Terra Marina.

Abel, an immigrant to South Africa, had come to Cape Town from Welkom after completing a government contract to work on the Free State goldfields.

Finding that a return to his craft in optics was going to be a lot more difficult than he had thought, he had a meeting at Three Anchor Bay with a man, Fred Jars, who said there was a living to be made out of diving for crayfish, along the Cape West Coast.

"So," says Abel, "I became a diver by trial and error, and the very next day – and for the next year or so – I learnt to make a living from one of the treasures of the sea.

"But the year 1962 marked the beginning of the end for commercial crayfish diving, when the government passed a law that had the effect of stopping the ever-swelling ranks of divers.

"Once more I seemed to be faced with a bleak future until, on a visit to Table Bay Harbour one day, I saw a weird-looking aberration of a barge being prepared to go to sea. I remember looking at a huge flow chart, fastened to the barge's steel framework, indicating that gravel was to be pumped up from the sea-bed at one end and diamonds recovered from the other.

"What I was looking at was a veritable jungle of steel and machinery. I had not seen anything like it before, and I tried to picture how this contraption was going to work. I was fascinated.

"To my surprise, I found on inquiring that there were vacancies for seamen/divers on this barge. I was ushered into the presence of marine superintendent Bob Greenshields who, after I had told him what I had been doing until now, told me I could start the very next day. I collected my belongings right away, and moved into the rather crowded accommodation on board.

"We sailed a few days later – on July 16, 1962, to be precise – and it was then that I first saw the boss of the whole operation, Sam Collins from Texas. He was on board with us, and my immediate impression of him was of a stockily built, always-on-the-roll leader who clearly commanded respect.

"Sam Collins came across as someone with unlimited energy, and with a deep knowledge of boats and of the men who go to sea in them – and also of the equipment they take with them. One also soon found that here was a man who looked after those who worked for him. He used to remark: 'Men who work hard also play hard – and they have to eat well!'

"A dented rust-bucket..."

"The dumb-barge in which we were now being towed was officially designated *CCC 77*. It was originally from the Collins Construction Company based at Port Lavaca in Texas, and it had clearly seen better days. A dented rust-bucket, it was still sturdy and seaworthy enough to make the long haul to Cape Town via the Persian Gulf.

"*CCC 77* had been refitted as a recovery plant – and now the hunt for diamonds was about to begin. I like to think of those early days as 'The Hunt', for very little was known about diamond deposits, and a lot depended on luck and on progress by trial and error.

"Heady days – and much hard and heavy work – lay ahead!

"To quote one of the guys on board: 'For us it is not a case of six days shalt thou labour and on the seventh shalt thou rest. Here, continuous production is supreme – and to get more than the others is everyone's dream'.

"Meaning, of course, that ours was a 24-hour-a-day, seven-days-a-week operation."

Hans Abel says when *Barge 77* arrived in MDC's concession area he was "a bit disappointed" to be assigned to the night shift. However, the first diamonds had been bagged and there was an air of excitement on board.

"We on the night shift were given explicit instructions to keep on pumping until morning, in the same spots that had been worked by the day shift, and to watch closely for any change in the weather.

"One night, when everyone else was tucked away in the sleeping quarters, we found that no more gravel was coming up, and no more diamonds recovered. We assumed there was a problem with the dredge pipe, and I embarked on my first-ever night dive.

"A bit hesitant at first, I found it not such a bad experience after all, with phosphorescent 'sparks' coming off my hands as I groped my way along the pipe, down into the pitch-blackness until I was able to feel into the nozzle to determine whether there was something blocking it.

"On returning to the surface we concluded that there was just no more gravel to be had in that position – and that was that.

"Now, I decided, was the time to give fate a little push. I suggested to Nicki Nienaber, the leading hand for the night, that the barge be moved just a little. He replied that I could do what I wanted – but that it would be on my head if anything went wrong.

"So warned, I carefully marked the cable position on our winch drum, and then let the brake slip so as to let out some more cable, after raising and lowering the dredge head into the new position.

"Gravel started coming up, and after about 15 more minutes in the sort-house I was looking at diamonds proudly produced by the night shift. About 6 am I started the anchor winch and moved the barge back to its original position.

"We had a tally of some 75 stones – many more than on the day before. We retired to have a well-earned rest.

"I was woken about 10 am and told that Sam Collins wanted to see me 'now-now'. With mixed feelings, I made my way to the boss's quarters in the 'Penthouse'.

"Collins looked at me long and hard, then said in his Southern drawl: 'What did you do last night?' I meekly replied that if I told him, he'd fire me.

'You'll be fired if you *don't* damn-well tell me!', he barked back.

"When I confessed that I'd moved the barge during the night, Collins told me to 'get your cotton-pickin ass back onto the winch' and to do exactly what I had done during the night.

"When I had done this, and the gravel reappeared, I returned to my bunk, happy to have gotten off so lightly.

"At last, production was starting to move!

"I should add that, being a trained miner, I realised that without ore-gravel no production was possible, and being on board a mining barge I further realised that one either had to move the dredge-pipe or the barge, or both, to find the gravel.

"And, as at that very early stage of our venture we were at the beginning of a learning curve, I was later jokingly given permission by Bill Webb to move the barge wherever I wanted to at night in order to find diamonds, 'as long as you wake me before we hit the beach!'"

Hans Abel says this first trip in *Barge 77* seemed to last forever. "I think it was close on two months. When we questioned Sammy about his promise that we would be able to return home in a week he retorted: 'Yea – but did I tell which goddam year?'

"When we finally reached Cape Town I was told I had been promoted to leading hand. Happily, I went on my way to play with the other toys that sailors play with – booze and girls!"

Later, when general manager Bill Webb introduced a bonus

scheme, as an incentive to the seaborne miners to work longer hours and to help boost production, Hans Abel became the biggest bonus-earner of them all.

Barge 77 at the mercy of a punishing sea

The sea-diamond miners soon found that the greatest threat to their offshore operations was the weather, and the sea conditions. Frequent gales and heavy swells made their work difficult and hazardous.

Hans Abel recalls that on July 1, 1963, while *Barge 77* was working in the Chameis Bay area, the crew swung the barge to face the weather, in a southerly gale that had started the day before.

"I had spent the day on the winch, watching for any emergencies that may arise. I was cold and wet when, at 1900 hours, Joe Horne came to take over the night shift. The wind had abated some, and I was glad to be able to take a hot shower and have supper, inside.

"When I made my way to the accommodation to turn in, I noticed that the wind had dropped almost completely, but the fog had come up and we were running into heavy swells.

"Although I was exhausted, sleep would not come. The barge started lurching around like a drunken sailor, and the port bow fair-lead made frightening noises every time it took the strain on the cable, as the barge dived into the next swell.

"It was about 3 am when one of our most experienced sailors – I think it was Sibley – got up and dressed in foul-weather gear. He had obviously had a notion of what was about to follow. Scarcely had he left the cabin when, with an almighty bang, the whole cabin disintegrated around us, and tons of ice-cold seawater poured over us.

"Double bunks suddenly became single bunks, lockers were smashed and as I got to my feet, miraculously unhurt, I noticed that where the door had been, only a hole in broken timber remained. A faint light at the end of the passage indicated the direction in which we must now make our escape.

"Once in the passage, I paused to make sure everyone was out, and then proceeded onto the upper deck, where I could look down on the anchor winch. Joe was there, but a puff of smoke from the winch exhaust told me something terrible was happening inside it, as the stress of the anchor cable forced it in the reverse direction.

"I watched, helpless, as the last of the wire was ripped off the drum and disappeared in a shower of sparks over the side. I remember Joe looking up as if to say: 'That's it!'

"A freak wave some 25 metres high must have hit the barge, and I knew we were now at the mercy of the sea. I realised I was wearing only underpants and getting very cold and very wet. A series of violent motions, and waves crashing over us, convinced me that I had to look for a safe place in case we turned turtle.

An "awesome silence..."

"What I still remember very clearly is the awesome silence between two breakers, and of sensing the tremendous weight of water suspended above us before it came crashing down to engulf everything.

"All the pumps and generators were by now silent, except for a little Deutz generator and one compressor that was still wailing away. I held onto a steel upright until once more my feet felt that the worst motions had ceased.

"When I dared peep over the side, I noticed dark patches in the white water – and realised we were on the rocks!"

Relieved that there was now solid ground under the barge, Hans Abel became more-than-ever aware of being extremely cold. He returned to what had been his cabin, and someone gave him a dry blanket. The crew were ordered to assemble in the penthouse, for a roll-call.

"To my amazement, a bottle of brandy was making the rounds, as we took stock of our situation. Fortunately, no-one was missing. My friend Fred Jars suffered a broken breast-bone, and that was the only casualty. The radio shack was wrecked and out of commission, but we could still communicate with our surveyors on shore, via the Storno, and tell them what had happened.

"The fog prevented any immediate attempt to leave the barge, but at least word of our stranding had got out – and we could all count ourselves lucky to be alive.

"Later, when the fog lifted, we were able to cast a heaving-line to our shore party, and then to get a sturdier rope to them. They attached this to their Jeep's winch. We then attached a running block to this rope and, with the shore party keeping the rope taut, we slid to safety, 'foefie-slide' style."

There were about 60 men on *Barge 77* at the time it hit the rocks, including Collins's 14-year-old son Sammy Junior from Texas and a Cape Town schoolboy friend Conrad Slabber, also 14. They all got ashore safely, using the emergency slide.

CDM at Oranjemund had meanwhile been notified of the grounding, and while the survivors awaited the arrival of their rescuers they made a big fire on the beach, to try and get themselves and their gear dry.

Landrovers sent to the wreck area by CDM arrived on the scene in the late afternoon. Sam Collins had meanwhile flown to Oranjemund from Cape Town and when Hans Abel, who had arrived in the first Landrover, met the Texan again he was clad only in an overall and gum-boots which were two sizes too big.

"I entered the officials' dining room like this after being ordered, as the first leading hand to arrive at Oranjemund, to report to Collins," he says.

"With him were the top-brass of CDM, all wearing suits and looking very important. I must have been a sight for sore eyes!

"Collins asked: 'Ain't you got nothin' else to wear?' I replied that everything else I had been able to rescue was soaking wet. Sammy turned to the manager of CDM and asked whether the general store could be opened for his crew.

"This was instantly agreed to, and we were all very appreciative of the concern shown for his men by our boss. To him, our welfare came before anything else – even before receiving a first-hand report on what had happened up there at Chameis Bay the night before – and he made sure we were all properly looked after, and clothed and fed.

"We shopped, showered, had supper and a few pots in the pub that night. Sibley, the sailor, kitted himself out in a dark blue suit and Crockett & Jones shoes. He had surely seen the chance of a lifetime.

"Next morning, we were put on a SA Airways scheduled flight at Alexander Bay, and we headed for home, thankful to be alive – and little the worse for an experience none of us will ever forget!"

More trouble – with 'Emerson K' badly holed

Little more than three weeks after the loss of *Barge 77*, Marine Diamonds suffered further misfortune when the tug *Emerson K* was badly holed near her stern by an uncharted rock pinnacle south of Port Nolloth.

The biggest gash was near the engine-room.

Prompt and effective action by the tug's divers saved the vessel from being flooded and sunk. The divers, D Botha of Bellville and E Manuel of Parow, had been hired for the *Emerson K*'s diamond-prospecting activities.

When the master of the vessel, Captain R Cranko, brought her astern and away from the pinnacle, the two divers rapidly donned their kit and went over the side, taking with them a mattress which they placed across the holes in the hull.

This action reduced the water-inflow enough to allow the tug's bilge pumps to cope. With the divers working on temporary repairs from inside the partly-flooded engine-room, the *Emerson K* slowly made her way under her own steam back to Cape Town, where she was drydocked for repairs before returning to the mining areas off the West Coast.

CHAPTER NINE

Back in the hunt – with bigger and better barges

WHILE the loss of *Barge 77* at Chameis Bay was a severe setback for the Marine Diamond Corporation, Sammy Collins with typical Texan get-up-and-go was back in the hunt for sea diamonds within three months of the disaster.

The barge had hardly run aground, with a financial loss to MDC of at least R250 000, when work was started by Globe Engineering Works in Cape Town on a second, bigger mining unit ordered by Collins.

He immediately had much of the wrecked barge's mining equipment salved and brought to Cape Town for installation on this new unit which came to be known as *Barge 111* – or the "One-Eleven".

In late 1963, this new barge became the biggest craft ever built in Cape Town – and it was put together in record time. About 120 welders and platers worked 24 hours a day and seven days a week, in shifts, to get the job done in the required time.

The contractors had to ransack sources on the Reef and in Durban for steel plates needed in a hurry – and the 700-ton barge was ready for side-launching in Table Bay Harbour only five weeks after work on it had begun.

Collins's technical assistant, Tom Kilgour, played a key role in the construction of "One-Eleven". He not only designed a bigger and better plant for the new barge; he was also one of those who worked night and day to get it on-site as quickly as possible.

Kilgour, born and brought up in Johannesburg, was one of Sam Collins's most-important "finds". A Wits University graduate with a master's degree in physical chemistry, he had worked for the Government Metallurgical Laboratories – later renamed the National Institute of Metallurgy. He had been on a research project for General Mining at Naboomspruit, in what was then known as the Transvaal, when he first heard about the Collins sea-diamond mining venture.

"While at Naboomspruit the consulting engineer, just back from Cape Town, told me that *Barge 77* was being fitted out by General Mining to mine diamonds off the West Coast. He also told me about Collins and his plans, and that sounded a lot more interesting than the research project I was involved in.

"I decided to go to Cape Town and ask Collins for a job. I did this, but the interview I had with him seemed to me to have been disastrous. He flung documents and questions at me, all to do with mining, and asked me for my opinion. I wasn't able to offer an opinion as I knew nothing of the particular subject-matter he was referring to.

"I was therefore surprised when he appointed me as a technical assistant. It turned out that what he wanted was someone with a university degree but without mining experience who could look after the technical aspects of his operation.

"In this way I became a mechanical engineer, charged with finding equipment that could operate and survive in a marine environment. General Mining had, in effect, taken a land-based mine and put it onto a barge. The result was that nothing worked, as most of the equipment could not take the motion at sea. It was therefore my task to adapt or design gear to operate efficiently afloat."

"Plenty of optimism..."

Meanwhile the "Lodestar" columnist of the *Financial Times* in London commented, in September 1963, that although sea-diamond operations off the coast of South West Africa had been temporarily halted by the loss of *Barge 77*, "plenty of optimism continues to radiate from the Marine Diamond Corporation camp".

"Lodestar" said Emerson Kailey, Sam Collins's right-hand man in Britain and Europe, had told him in London that the building of new vessels was well advanced and that diamond production should be resumed very soon.

"It is hoped to make an introduction this month to the London market – and possibly to the Johannesburg market as well – of the one-shilling shares in Sea Diamonds, the company which holds Collins's stake of more than 40 percent in MDC. Of the 14,3 million issued shares in Sea Diamonds, something like 0.5 million will be made available in London.

"Just what sort of price basis they will command is anybody's guess at the moment, but it could possibly be upwards of 20s – at which level they are currently reported to be changing hands in unofficial dealings in Cape Town. Such a price puts a capitalisation of £14,3 million (R28,6 million on the exchange rate at the time) on Sea Diamonds itself.

"Much as one must admire the Collins tenacity, one cannot help feeling that such a huge capitalisation is high for a pioneer mining enterprise which is at the same time unique and also in its infancy. This becomes especially apparent when one considers that General Mining, a long-established finance house, is only capitalised in the market at a mere £21 million (R42 million).

With the introduction of aircraft to service the mining rigs, the handsome Schipa *became Sammy's personal "yacht".*

"This company's stake in Marine Diamonds forms a relatively small part of its widespread interests."

"To hell with shipwreck leave..."

Joe Horne recalls that while the survivors of *Barge 77* were entitled under international law to seven days "shipwreck leave", Collins was not interested.

"To hell with shipwreck leave, he told us; there was work to be done right away, and he wanted some of us back at Chameis Bay without delay to get the gear off the wrecked barge.

"There were 10 or 11 of us in the salvage party, and for three weeks we lived in a shack (Fort Reef) on the shore at Chameis Bay that had been used by our party of surveyors. We worked non-stop for those three weeks, stripping the barge of everything that could be used again. Our days would start about 4 am, so that we could get the heavier work done before the heat of the day got to us.

"We lived off the stores still on board the wreck. One guy's staple diet for three weeks was Cape cherries; another lived on pink salmon!

"We had to conserve our small supply of fresh water, so none of us washed for three weeks – and we all burnt black, working out in the open all the time. We were not exactly nice people to be near while we were up there.

"When all the movable gear had been taken off the wreck, it was transported to Luderitz and from there it was shipped by coaster to Cape Town, to be fitted to the new barge that was under construction here."

Barge 77 had proved too short for the wave-frequencies off the West Coast, slamming into the swells and overstraining its moorings. The new *Barge 111* was therefore made 12 metres longer, to enable it to ride the prevailing swells more comfortably. Two additional seaward mooring anchors were installed (making six anchors in all), together with four 2 000-hp hydro-jets to relieve the strain on the anchors in heavy weather.

Where *Barge 77* had carried a single airlift for sucking diamonds from the sea, "One-Eleven" was equipped with two 30-cm (12-inch) lifts, each with its own scalping and washing screen circuits, and a single heavy media plant. Accommodation for 65 persons was provided.

And where the "77" had had a squared-off, flat-roofed deck-house at the after end to accommodate its crew, the new barge – and those that followed it – had a dome-shaped structure, reinforced to withstand the onslaught of the sea.

Apart from being longer, *Barge 111* was also wider, and with all these improvements and refinements in its design, this Cape-built craft went on to gain a reputation as a "dry boat", ideal for the conditions in which it had to operate.

Ultimately, it also proved to be the most successful of all Collins's marine mining units, producing 5 000 to 7 000 carats a month in the Chameis Bay area, in about 20 metres of water – the "magic" depth for wherever diamonds were to be found off the West Coast.

While in later mining units Collins was to try other systems for recovering diamonds, including jet-lifts and centrifugal pumps, the airlift – a system that had been discarded as archaic by the rest of the dredging world – proved in the long run to be the most efficient method of mining gemstones at sea.

This was because it was less prone to constriction problems and blockages than other systems and could, with no moving parts and simply by injecting air at the bottom to cause suction, draw materials of a larger size all the way up the pipe, unimpeded.

The airlift system is still being used effectively in sea-diamond mining at the time of writing.

'Colpontoon' and 'Diamantkus'

By September 1963 "One-Eleven" was ready to sail, and had been towed on-site and commissioned at Chameis Bay by November. Within sight of the wrecked "77" – the twisted hulk of which remains on the rocks to this day – the new barge was soon able to take up where Collins's first mining plant had left off.

And while the new barge continued to dredge diamonds from the floor of the sea, two more vessels were being converted in Cape Town, for use as mining units.

These were the *Colpontoon*, with four times the capacity of "One-Eleven", and the 4 880-ton former US Navy tank landing craft *APB 45*, which Collins had acquired in America in mid-1963 as an addition to his growing fleet.

Altered beyond recognition for her new and glittering role, the former warship was dedicated by the then Administrator of South West Africa, Daan Viljoen, and was formally renamed *Diamantkus* by

The third mining barge, Colpontoon, *leaves Table Bay for SWA. Note the protective, dome-shaped crew accommodation.*

Gigi Collins at a ceremony attended by many invited VIPs in Table Bay Harbour in October 1963.

Now the world's largest seagoing diamond-mining plant, *Diamantkus* became operational in the Marine Diamonds concession area off SWA in January 1964, under command of Captain George Foulis, who had delivered the vessel from the USA.

When the ungainly-looking ship sailed from the Duncan Dock in Cape Town after her conversion, the scene at her departure was said to have been "like mailship sailing day".

In a gala atmosphere, with the champagne flowing, *Diamantkus* was seen off by hundreds of cheering and waving relatives, friends and holidaymakers, and the *Cape Times* reported: "Standing unnoticed among the crowd – and no doubt with thoughts and hopes flashing across his active mind – was the Texan 'sea-diamond king' himself, Mr Sam Collins."

The self-propelled *Diamantkus* had five sea anchors, accommodation for 200, three 46-cm and three 41-cm airlifts and two complete recovery circuits. A crossfeed arrangement allowed head-feed to be diverted from one circuit to the other, to allow for repairs to be carried out.

The vessel could handle 300 tons of gravel (roughly 200 cubic metres) an hour – ten times as much as *Barge 111*.

After its conversion the *Diamantkus* looked like something out of an Emmett cartoon, its decks cluttered with a tangle of pipes, valves, winches, conveyors and other mining structures. "Going from fore to aft was like undergoing an assault course," says Tom Kilgour.

The 1 621-ton *Colpontoon*, originally a tidal landing pontoon, was fitted out for diamond recovery in September 1964. It was 80 metres long with a beam of 15 m and was fitted with six mooring anchors and one 46-cm and one 41-cm dredge pump, both of which could be replaced with airlifts if required. As in *Diamantkus*, two complete recovery circuits were installed. The *Colpontoon* had accommodation for more than 80 men.

The barge's hull had been built by the Fairfield Shipbuilding and Engineering Company in Monmouthshire, England, and her mine processing plant was fitted in Cape Town.

The October 1964 issue of *SA Shipping News and Fishing Industry Review* reported that the Marine Diamond Corporation had, perhaps, become the largest single private enterprise employer in the Cape Town docks, granting contracts not only to the Globe Engineering Works but also to a number of other local firms.

Because of a shortage of artisans, the journal said, MDC had formed its own engineering department. "It was the men of this department who completed most of the work on the *Colpontoon*, although the welded shell for the mining plant was again built by Globe."

Superintendent of the construction engineers was the MDC's Ronal Wilsenach, a technician with experience gained on the Williamson mines in Tanganyika (now Tanzania) and in South West Africa.

Wilsenach was quoted at the time as saying the company's artisans had been working 24 hours a day, in two shifts, for about three months on the *Colpontoon* project.

This US ex-tank landing craft, APB-45, was acquired and converted by Collins in 1963, and renamed Diamantkus.

After conversion in Cape Town into a mining unit, this is how Diamantkus *looked when she sailed for SWA in 1963.*

"A feather in the cap of Cape engineering firms"

This conversion, from a dumb-barge into a fully-equipped floating mine, jammed with machinery and plant of every description, was seen as another feather in the cap of the Cape Town engineering firms responsible for this pioneering work.

Meanwhile the journal *Diamond News and SA Jeweller* carried a major article in its May 1963 edition in which it was stated that "South West Africa is experiencing the biggest scramble for mineral wealth in its history".

Mining groups and investors from many countries were spending large sums on concessions and on intensive prospecting, the journal said.

"In addition to the diamonds, the value of which last year totalled more than R34 million, they are seeking gold, oil, copper, coal, tin and a variety of other valuable minerals such as tantalite and vanadium. Last year 424 prospecting licences were issued; 86 more than in 1961.

"At present most interest is focused on the hunt for sea diamonds – begun by Mr Sam Collins and now joined by a group of South African finance houses.

"SWA's two most important existing mining concerns are the Consolidated Diamond Mines, working at Oranjemund, and the Tsumeb Corporation Limited, whose output of copper and other minerals rose by more than R1 million last year, to a total of R15 702 163."

In January 1964, before the *Colpontoon* came on the scene, diamond production from the SWA seabed by *Barge 111* and the *Diamantkus* was reported to be "hitting new peaks".

The combined production on one day, in the week preceding the report, had been 1 397 carats, and MDC experts had estimated that the whole organisation needed only 500 carats a day to break even.

The *Financial Mail* of January 31, 1964, said the *Diamantkus* (target production 2 500 carats a day) was, after initial teething troubles, edging its daily figure up towards one thousand.

"It is still having production problems and only two or three of the six airlifts have been used in the past week. But its total of 3 100 carats for the first two weeks of production has been eclipsed by the figures for the last week."

The journal said the latest figures put the *Diamantkus* production at close to 4 000 carats in under a week, and *Barge 111* was sucking up far more diamonds than had been expected, at nearly 2 500 carats. The total for the six days was over 6 000.

A Sea Diamond Corporation report published in London in April 1964 said prospecting off the SWA coast at that stage had resulted in the discovery of two diamond-bearing areas – the so-called Chameis Area, including deposits seaward of Chameis Head and South Rock, and the Plum Pudding Area, between Plum Pudding Island and Noah's Ark.

"These areas alone are believed adequate to sustain full-scale mining activity for many years," the report added.

"In the Chameis Area, including the so-called 'Surf Zone' in which *Barge 111* is now operating, it is estimated that the different sediment bodies that have to date yielded diamonds cover a total area of some 8,8 million square metres. The potential of an additional 3,8 million sq-m constituting the 'Surf Zone' – ie the area stretching from low-water mark 300 metres out to sea – has not yet been fully investigated due to shallow water-depths and heavy surf.

"It is believed that the diamondiferous gravels would extend into this untested 'Surf Zone'.

"In the Plum Pudding Area, indications are that diamondiferous deposits cover an estimated area of 3,25 million square metres seaward of the 'Surf Zone'.

"Yields from prospecting work performed by the *Emerson K* in this area are comparable to those found in the Chameis Area, and it is felt that the potential per square metre of this area may be no less than the Chameis deposits."

CHAPTER TEN

Aircraft acquired but more setbacks afloat

THERE were eventually three floating diamond-recovery units operating at one time in Sam Collins's concession – all of them mining 24 hours a day.

Crew members at that stage were working 20-day, 12-hour shifts and then returning to Cape Town for a week's rest before resuming the work-cycle. This posed the MDC with crew-change logistical problems, and the acquisition of suitable aircraft became a priority.

Initially the corporation's motor yacht *Schipa* was used to ferry the barge crews, but this took time and caused low morale among the sea-diamond miners, many of whom became violently seasick in the lively little vessel on the long haul between Luderitz and Cape Town.

The *Schipa*, originally owned by JW "Bill" Mitchell of Hout Bay, one of Collins's co-directors in the Sea Diamond Corporation, was now jointly owned by Mitchell and Collins. Their partnership chartered the vessel to Marine Diamonds, and this became a lucrative venture for both partners, all maintenance and repair costs – and improvements – being charged to MDC's account.

Schipa's one-time chief officer, or mate, Eddie Redgrave, remembers the trip from Cape Town to Luderitz taking 36 hours – "and it was always the middle of the night when we got there.

"As if that wasn't bad enough; the crew-change people had to be ferried back and forth between *Schipa* and the rig in a little dinghy with an outboard motor. Jock Carr from Scotland used to drive this dinghy, and in the darkness we'd see him bobbing up and down alongside us with people carefully having to time their jumps in and out of his tiny boat. It says a lot for him that we didn't have any drownings!"

This sick-making sea shuttle resulted in a high turnover of mining personnel. Many of the men simply refused to endure more turbulent, tummy-churning trips along the West Coast in the *Schipa*.

In dealing with this problem Marine Diamonds purchased or chartered a small fleet of fixed-wing aircraft, including three twin-engined Douglas DC3s (Dakotas) which Collins named *Cottonpicker*, *Ruffneck* and *Hossfly*.

The first Dakota, *Cottonpicker*, had been rescued out of a snowbank in Alaska. After repairs, the 27-year-old aircraft was flown to Cape Town by an associate of Sam Collins, Bob Fridell of Victoria, Texas.

Fridell ran Collins's office in Port Lavaca at the time. Later, he and his wife Frances, on their first holiday in many years, were killed when their light aircraft crashed into the mountains of Colorado.

Collins's small fleet of ferry aircraft was eventually expanded with the purchase of a four-engined DC4 Skymaster, and a twin-engined Beechcraft Bonanza was acquired for the personal use of Collins and his executives.

In 1967 a Beechcraft Baron that had also been used by the MDC landed on the Luderitz airstrip with its undercarriage up. No-one was hurt, but the aircraft was a write-off.

Collins's first chief pilot was Bill van Heerden who, with Hugh Pharoah, started the MDC's flying operation. The other captains apart from Van Heerden were Ainsley "Cookie" Cooke, Golly Oakes, Piet Albertyn and Johnny Human. The co-pilots were Pharoah, Carl Middle and Laury Gianni.

The engineers who accompanied the pilots for "field handling" were Brian Johnson, Bob Illingworth and George Gibbs.

Captain Hugh Pharoah, who went on to fly wide-bodied aircraft for SA Airways, Air Mauritius and Singapore Airlines, recalls that the first aircraft acquired by Marine Diamonds was a Cessna 182, either bought from or loaned by one of SWA's best-known personalities, Dr Peter le Riche.

"*Cottonpicker*, our next aircraft and our first Dakota, was an original DC3 – the DC3A – as distinct from the converted war-surplus C47. It had Wright Cyclone engines and a bungy chord undercarriage damping system, and it was the only aircraft I knew where, if you put down a lousy landing, it was still bouncing up and down by the time you set the parking brake at the terminal building!

"The only thing missing was the piano..."

"It came to us from Mohawk Airlines in its 'Gaslight Service' livery, complete with its old-time saloon fittings and velvet drapes. The only thing missing was the piano which, I gathered, had been in it previously.

"At first Sammy used *Cottonpicker* as his personal aircraft, and we had some very interesting flights to Windhoek in it, while he was finalising his mining concessions.

"I recall one incident when I was doing the flying – the old girl not having been fitted with an auto-pilot. Bill van Heerden had gone to join the festivities at the back, and had shut the cockpit door. Suddenly, the trim of the aircraft began to do strange things, becoming progressively very tail-heavy till I could barely hold it, then equally suddenly becoming very nose-heavy.

"I struggled to fly the thing in this manner for a while, not knowing what was happening behind the closed door, and not having a PA (public address) system with which to communicate with the others on board.

"I eventually throttled both motors back and that succeeded in getting the attention of those at the back, and bringing Van and others running to the cockpit.

"Then I saw that they had stripped the seats out, piled them against the side and were playing bok-bok against the rear bulkhead! It was a very successful flight for the passengers, judging by the way we later poured them out on the apron at Windhoek!

"We then acquired another two DC3s from WENELA (the Witwaters and mines labour recruiting organisation) and these were both maximised aircraft. Both very Spartan, they had no heaters or air-conditioning, so in the winter we froze and in summer we baked. These aircraft were used purely for personnel transfer to and from the rigs.

"Shortly after I joined the company we got rid of the Cessna and replaced it with a Twin Bonanza. This aircraft was the company commuter, used by all the top management including Sammy Collins. I got to do just about all the flying of this aircraft, and so virtually became Sammy's personal pilot.

"The only drawback to this was that I would quite often get a call in the early hours of the morning from Sammy or his colleagues from a nightclub on the Cape Town Foreshore saying they wanted to go somewhere after the party. This meant having to go out to the airport to get the plane ready.

"I subsequently got in the habit of always leaving the tanks full and having the aircraft parked right at the front of the hangar."

Hugh Pharoah recalls that after he had left Marine Diamonds, *Cottonpicker* was put on her nose while doing a compass swing, and that when the old Dakota was stripped down for repairs, so much corrosion was found in the wings that the aircraft had to be scrapped.

"She ended up very ignominiously on her belly as the Cape Aero Club's clubhouse at DF Malan Airport (now Cape Town International Airport) – a great pity, as Sammy had told me he intended donating it to the Smithsonian Institution in Washington DC.

"About this time, Marine Diamonds rationalised their operation and acquired a DC4 (Skymaster) and used the old DC3s only as overflow transport. One of them – ZS-DBP, named *Hossfly* – was subsequently operated by National Airways and crashed just after take-off at Rand Airport, Germiston, killing all on board."

Eddie Redgrave, who after his stint in the *Schipa* worked on the barges as a marine superintendent and later as a mining superintendent, says a memorable feature of the MDC Skymaster was that its starboard outer engine often failed to self-start on the desert landing-strips.

"The way they used to get around that little problem was to get the engine started with a length of nylon rope. The one end of this rope would be coiled round the propeller-boss, and the other end would be tied to the back of a Landrover.

"The Landrover would then drive off smartly at right-angles to the aircraft, spinning the propeller in the process and getting the engine started. It was like starting an outboard motor every time.

"Our lifeline was therefore that piece of nylon rope. They always carried it around with them, up in the desert. How the devil that plane was ever allowed to take off from the airport at Cape Town, I'll never know!"

Thanks to Sammy, Luderitz gets a new airport

The Collins group decided to provide Luderitz with a new airport, as the airstrip which had served the desert port for many years was too short for big aircraft. It was also sited near a salt lake, and was prone to waterlogging after rain.

While the MDC paid the R30 000 required for construction of the new runway, about 3 km north of the original strip, the Luderitz municipality paid for the R6 000 airport building and for certain road deviations in the area.

After landing at Luderitz the barge crews were taken by a converted Norwegian fiord ferry, *Marina*, to the mining area where they were transferred by dinghy – often at night and often in heavy fog and with high seas running.

When acquired by Collins in 1964, the relatively new *Marina*, fitted with twin diesels and capable of carrying 300 day-passengers, was reported to be the only passenger ship on the South African register.

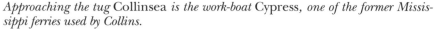

Approaching the tug Collinsea *is the work-boat* Cypress, *one of the former Mississippi ferries used by Collins.*

The MDC's senior master and marine superintendent, Captain George Foulis, was given command of the vessel and thus became the first master of South Africa's first registered "passenger liner".

Two converted whalecatchers, the *Aunty Lil* – nickname of a popular member of Collins's office staff in Cape Town – and the *Oom Kappie*, named after parliamentarian Kappie Strydom, were at one time also used for transportation of crew and supplies.

Other auxiliary vessels in the MDC fleet, used in a variety of roles from time to time, were the *Cypress*, the *Chan*, the *Willow*, the *Cortes*, the *Chameis*, the *Mashona Coast*, the *McLaughlin*, the *HC Seymour* and the *Yolanda*.

Several of these vessels – all about 20 metres long – were brought across from the United States towards the end of 1962, to serve as personnel and supply carriers, and as general support craft. They were typical of support vessels operating on the Mississippi River and although not of a design suited to the rough and exposed seas off the diamond coast, they served the MDC well for several years.

One of them, the *HC Seymour*, sank early in 1963 on her way to Durban. Another, the *Cypress*, which had twin screws and was highly manoeuvrable, was used for geophysical work.

By 1965, when De Beers took over, chartered helicopters had been brought into use, to shuttle crewmen between the barges and the airstrips at Luderitz and Chameis Bay.

In December 1964 the *Cape Times* quoted Sam Collins as saying that the *Colpontoon* had shown "excellent results" since it had come into service three months earlier. A total of 21 859 carats had been recovered from the sea by the barge in that short period, he was reported to have said.

The barge, he added, had been let to the MDC at a rental of R40 000 a month, and the firm had the option to buy it outright after six years, at half its price of construction.

Collins said *Colpontoon* had a net return of R20 a carat. Out of this, the operating cost had to be met, which amounted to half, or 2 000 carats. It needed another 2 000 carats of diamonds to be found each month for the rent.

Figures released by Collins showed that 3 655 carats had been mined by the *Colpontoon* for the last nine days of September, 10 858 for October and 10 346 for November.

He had recently put R800 000 of his own money into the company, he added, in addition to R400 000 from elsewhere. This was necessary to go ahead with expansion plans.

Collins described as "nonsense" a recent rumour that 90% of the stones recovered had been industrial diamonds.

"In fact, more than 99% are gem diamonds. We may get one industrial diamond in 1 000 stones. Considering that to mine two-and-a-

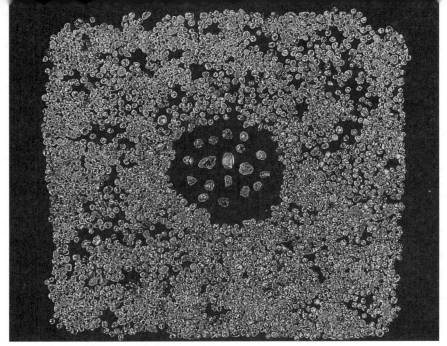

A haul of 1100 carats of gem diamonds taken from the sea in 6 days. The larger stones (centre) weigh 2 to 5 carats.

half carats you have to process one ton, this is neglible."

So confident was Collins in the future of his sea-diamond mining venture in 1964, that he announced plans to build a R10 million skyscraper – "the biggest and tallest in Africa" – on the Cape Town Foreshore.

This, he said, was to be the headquarters of his organisation in Southern Africa. The building had already been designed, and drilling had started on a Foreshore site to determine how high the skyscraper could be built.

Collins also announced plans to build a R1 million engineering plant at Epping, an industrial area within the municipality of Cape Town.

The 'Colpontoon'/'Collinstar' disaster

In February 1965, when a take-over by De Beers appeared imminent, Marine Diamonds suffered a severe blow to its operations when the *Colpontoon* dragged anchors and was driven ashore in an exceptionally heavy sea.

In attempting to save the massive barge, the former British tug *Collinstar* capsized and sank, with the loss of six lives. Some survivors among the 13-member crew, sensing imminent disaster, managed to jump onto the barge as the tug bumped into it. Others managed to struggle ashore.

The master of the tug, Captain Karl Gobel, a 29-year-old German immigrant living in Cape Town, was one of three men washed up, drowned or near-drowned, on the beach. He was the only one still registering a weak pulse, and for more than an hour members of the shore party tried in vain to revive him, with mouth-to-mouth resuscitation.

In their rescue and resuscitation attempts, the shore party lit the beach in the gathering dark, with lights from their vehicles.

A surveyor employed by Marine Diamonds, Jurie Terblanche, emerged as one of the heroes of the rescue attempts. He swam through boiling surf to the stranded *Colpontoon* to bring back its medical officer who was needed on the beach. The two men were pulled ashore in an inflatable rubber raft.

Colpontoon was later refloated by the biggest tug in the MDC fleet, the 4 500 shaft horse-power *Collinsea* – the former Dutch salvage tug *Zwarte Zee*. Commanded by Captain Graham Harker, the big tug towed the damaged barge back to Cape Town for repairs.

Plant manager Graham Godfrey, who witnessed the *Collinstar* tragedy, later described how the tug was broadside to the sea when hit by a gigantic wave. "As I looked back at her, the tug flipped over like an empty beer-can, her twin screws up in the air."

Captain O E "Okkie" Grapow, a director of Safmarine at the time of writing, was mate (later master) of the *Collinsea* when the *Colpontoon* was driven ashore.

'I nearly lost my legs in that affair, when a taut wire we had got across to the *Colpontoon* parted in the swell, and backlashed. After we had towed the barge back to Cape Town I was put off on four months sick leave.

'When *Colpontoon* started dragging anchors in a heavy sea, the Americans wanted us to go through the outer line of surf to where the barge was practically aground already. We refused. The tug was too big and deep to go so close in, beyond the first line of surf.

After dragging anchors in a heavy swell, Colpontoon *lands on the beach at Chameis Bay, in February 1965.*

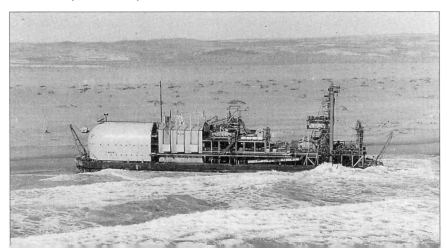

"The captain of the *Collinstar*, a smaller tug, thought it looked all right and he ventured in, between the two lines of surf where there was a sort of flat patch. He might have made it with a smaller and more manoeuvreable tug.

"But the heaving *Collinstar* hit a corner of the barge, lost stability and buoyancy and suddenly flipped over in the surf.

"Then suddenly she was gone..."

"I think what happened was that the tug was holed in the collision and the engine-room got flooded. I remember looking out of a window on the bridge and seeing the *Collinstar* approaching the barge. One minute she was there, then suddenly she was gone.

"The tug's wheelhouse was torn off. We looked hard for survivors, but found none. The captain and chief engineer were among the six lost in that affair.

(The chief engineer was JW Wisboom-Verstegen of Simon's Town. Like Captain Gobel, he was only 29).

"Although it was a sunny day, with little wind, the sea was wild. Before the accident we in the *Collinsea* had been doing an echo-sounder trace of the waves, and we had estimated the swell at 50 ft (15 metres).

"And just before the loss of the *Collinstar* I remember seeing a small line-running boat called the *Cortes* heading towards us with a line from the *Colpontoon*. I went down to the after-deck to get this line, and I can still picture the little *Cortes* crashing through a huge wall of water as it headed towards us. It was awesome.

"To give some idea of the immensity of the seas in that area we

In the bid to refloat Colpontoon, *the Emerson K (left) stands by as a helicopter tries to get a line to* Collinsea.

MDC's biggest tug, the Collinsea, *which hauled the stranded* Colpontoon *off the beach at Chameis Bay.*

could, in the distance, see the wreck of *Barge 77* at Panther Head at the southern end of Chameis Bay. In 1963 it had been lifted bodily and dumped on the rocks, so high up that it never got really wet again.

"In the exceptional conditions on the day the *Colpontoon* ran aground, the remains of the '77' caught a bit of spray – but that was all!"

When Sam Collins, then in Cape Town, was notified of the disaster he immediately arranged for a company aircraft to fly him to the seamining area. He was able to get on board the grounded *Colpontoon* at low tide, from the sandy beach.

It was typical of him that he should want to be on the spot whenever anything went wrong.

After intensified efforts to free the *Colpontoon* from the desert sands had eventually paid off, a financial writer commented: "Collins's success in this salvage operation should be quite a hit with the insurance writers, whose hearts must have missed a few beats at the news of the second major sea mishap.

"Loss of lives might deter some future crew members, but admittedly this danger is present in any mining project, below ground or at sea.

"The accident comes at a delicate stage on the company front, with De Beers due to decide on taking part in Marine Diamond Corporation equity by the end of March."

In a letter dated February 18, 1965, and directed to all personnel of Marine Diamond Corporation, Sam Collins wrote:

"The tragic death of Captain Gobel and his crew members is certainly to be regretted by each and every one of us. I understand the personnel of the mining units are taking up collections for the survivors of this tragic accident. This is indeed commendable.

"Captain Gobel and his men lost their lives trying to save the *Colpontoon* and its crew of eighty people. They all knew that their lives were at stake and they sacrificed them to try to save the greater number of men.

"All of you should well remember the courageous actions of these people. As has been quoted: 'Greater love hath no man than to give his life for his fellow workmen'.

"Captain Gobel and his crew unselfishly gave their lives and went down with their ship, upholding the age-old tradition of the sea. We shall always remember them as heroes of a tragic accident."

Soon after writing this letter, and at a ceremony in his offices on the Cape Town Foreshore, Collins presented gold cufflinks to employees of his MDC for their "valorous service" during the grounding of the *Colpontoon* and the loss of the *Collinstar* at Chameis Bay.

In making the awards, Collins said the men receiving them had risked their lives attempting to save others, and that they had given "outstanding service beyond the normal call of duty".

One of the six men lost in the *Collinstar* tragedy was the tug's 38-year-old chief officer, Alan Smith of Kenilworth, Cape. The father of five children, he and his wife Pat had recently arrived in South Africa from Malawi, where Captain Smith had spent four years in the Lake Malawi service.

In February 1965 Pat, originally from York in England, to which she returned, sent a letter to Captain Graham Harker of the *Collinsea*:

"I write to thank you and your fellow officers of the *Collinsea* for the sympathy and flowers you sent for Alan.

"I know how you all must feel at the loss of fellow seafarers, and I can do no better than to quote a passage which is most appropriate to my husband:

'His life was gentle, and the elements so mix'd in him that Nature could stand up and say to all the world: This was a Man.'

"Our children have a lot to live up to, in having been blessed with a

The ill-fated tug Collinstar, *which overturned with the loss of six lives, while trying to save the* Colpontoon.

Diamantkus *at work off the SWA coast. The vessel proved uneconomical, and ended up as a Defence Force target.*

fine father, and I hope to be able to bring them up to be proud of his memory, as I am so proud and happy to have been his wife."

A close call in 'Diamantkus'

Jock Carr, a former leading stoker in the Royal Navy and a coalminer at Fife in Scotland, held positions first as foreman and then as marine superintendent and mining superintendent on MDC barges.

Signed on in 1963 as an employee of Collins Submarine Pipelines Africa (Pty) Ltd, he was seconded to Marine Diamonds, and was on board *Diamantkus* in 1965 when the one-time tank landing craft came close to sharing the fate of *Barge 77* and *Colpontoon*.

"We had been operating close inshore, only about quarter of a mile off the beach. One day, a record number of diamonds for a single day-shift – about 5 000 – were brought up and Sammy Collins radioed from Cape Town: 'Well done! Everybody have a drink on me!'

"I was a foreman in *Diamantkus* then, and by the time I got to the bar, on the upper deck, I could see that just about everyone was sozzled already. I ordered a beer, and I had only taken a few sips when the mining superintendent came dashing in and shouted 'Jock, I need you!'

Jock Hendry was also there, so I asked which Jock he wanted. Jock Hendry was known as Big Jock and I was called Wee Jock. The mining superintendent replied 'I need both of you! – the deep-stern anchor cable has broken and our starboard-deep's dragging'.

"We used five anchors in *Diamantkus* – two sterns, a deep-stern and two bow anchors.

"When we got outside, I could see that we were indeed in trouble – with the deep-stern cable parted as a result of the pounding we were taking in a heavy sea. And the shore looked ominously closer than it had before. We were obviously dragging towards the beach.

"With broken cable hanging over the side, close to the propellers, *Diamantkus* was unable to use her engines to get out of this under her own power. The tug *Collinsea* was trying to get a line across to us, but without success. I then managed to get a heaving-line across to the tug, and a towline was connected. We were right in the breakers by then, and beginning to touch bottom.

"As the *Collinsea* took the strain and managed slowly to pull us seaward, we let the remaining anchors go. We were towed to a safe position offshore and we had to wait there for a week, before the weather calmed sufficiently for work-boats to go in and retrieve our anchors and cables, all with marker-buoys attached to them.

"Then, having got all this gear back on board *Diamantkus*, we returned to the exact co-ordinates at which we had brought up that record intake of diamonds – only to find that we could not locate the exact spot. There was absolutely nothing there!"

Crayfish for Africa!

Jock Carr says that on another occasion when *Diamantkus* had her pumps operating in a diamondiferous area, hundreds of crayfish suddenly came up the suction pipes.

"We were used to the idea of all sorts of strange objects coming up through the pipes – once we even sucked up an empty gas-bottle – but these crayfish were something else.

"All of a sudden they appeared on all the screens. Then they went across to the conveyor belts – and back over the side. There must have been a migration of crayfish; they were all over the place – and in vast numbers.

"We quickly started pulling them off the screens. The boilermakers cut the tops off empty oil-drums and the crayfish were thrown into them, boiled and then taken to the refrigerators.

"We couldn't possibly use all the crayfish ourselves, so we had them distributed to the entire fleet, with our compliments. And at the next crew-change, everyone going back to Cape Town had a bundle tucked under an arm when he stepped off the plane – in addition to the usual suitcase.

"Our security officer, Rex Redelinghuys, just about went crazy when he saw this. 'What the hell is going on here?' he demanded to know. 'There could be diamonds in every one of those parcels, for all I know!'

"Well, someone who had had the foresight and presence of mind to bring a big parcel of crayfish for our security chief responded by saying 'and these are for you, Rex!'

"We were all able to dine on West Coast crayfish at our homes that night – but there was a clamp-down on the taking ashore of parcels of crayfish, biltong and bokkems (Cape dried-and-salted mullet).

"Everyone had by then heard of the chap at Oranjemund who asked his chum who was going on leave to deliver a parcel of biltong to a certain address in his hometown. When, at the security gate, this chap was asked if he was carrying biltong, he said yes and would the security officer like some? The security officer accepted a slice – and nearly broke a tooth on an uncut diamond!"

Another story about *Diamantkus* that Jock Carr recounts concerns one of Collins's top managers who, under prolonged stress caused by onerous working conditions and long hours on duty, had developed a drinking problem.

"On one memorable occasion when this man visited *Diamantkus*, he somehow got hold of a firearm and suddenly started shooting at the penthouse windows. He just went berserk, and there was glass flying all over the place.

"Instead of firing the man, Sammy had him sent abroad for specialised treatment of his stress and drinking problems – and reinstated him when the course of treatment had been completed. This was typical of the care and the kindess that Sammy showed his staff. Rather than punish a person in circumstances such as this, he would help him overcome his problem."

Loss of the 'Emerson K'

The year 1965 was a disastrous one for Collins and Marine Diamonds. The *Financial Mail* commented: "The flamboyant style of MDC's pioneer, Sam Collins, are over. They have been exchanged for quiet domesticity under the comfortable mantle of groups like De Beers and Bonuskor."

The journal said the latest reports from the undersea diamond front had been disappointing, and it referred to "the marine mishaps which overtake Mr Collins with unlucky frequency in his unending struggle with the cruel sea".

The latest blow in MDC's saga of mishaps came on August 9, 1965, when the *Emerson K*, Collins's first prospecting vessel, rolled on its side and sank in the mouth of the Robinson drydock in Table Bay Harbour.

The tug, which had been undergoing repairs and a refit elsewhere in the harbour, had been patched up – but not made sufficiently watertight – for the short journey from the Duncan Dock to the adjacent Alfred Basin.

She was being nudged by a harbour tug into the entrance of the drydock when, with water rushing in through rivet-holes in loosely fitted hull-plates which were to have been rivetted in place in drydock, she suddenly heeled over on her port side.

As the big tug came to rest on the sill of the drydock, completely blocking the entrance, the 24 men on board her – the crew and ship-repairers – scrambled to safety along her upturned hull.

And once again, one of the first persons to arrive on the the scene was the Marine Diamonds boss, Sam Collins, who immediately took personal charge of operations.

The half-submerged hulk posed an unusually complicated salvage problem for Sammy. Not only was it a question of righting and refloating the vessel; this also had to be done in such a way as not to damage the sill on which it rested.

The sill is the underwater seating for the caisson that closes the entrance of the drydock, and any damage to it would have meant the drydock would be out of commission for months, necessitating the construction of a coffer dam and the pumping out of water while repairs were effected.

Salvage operations began with the rigging of ropes and steel cables to the tug. These were used to pull on the vessel's starboard side, in efforts to raise her onto an even keel.

The next step was the removal of as much as possible of her above-deck weight. This included diamond-mining platforms, vibrating screens and jigs weighing some 60 tons.

In cold and difficult conditions, divers set about inserting water bleed lines and air purge lines into the various tanks and compartments, and sealing these with makeshift wooden manhole covers and doors. These had to be bolted into position under water, sufficiently watertight to allow the pumping out of water at a faster rate than it could leak in.

The battle was complicated by the tide, which still rose above the level of deck openings.

After an unsuccessful attempt was made to stem the flow of water into deck openings by stacking bags of sawdust, steel coffer dams were built, effectively preventing the inflow of the sea.

Slowly, *Emerson K* began to float on her side. Six tractors now came into action to get the tug onto an even keel. Four of the two-inch steel cables used, each with a breaking strain of 200 tons, had to be

Collins's tug Emerson K *overturns in the entrance to the Robinson drydock, Cape Town. She was salvaged and scrapped.*

A close-up of the bows of Emerson K *as the overturned tug settled on her port side, on the sill of the drydock.*

replaced during this operation.

After 12 days semi-submerged, the mud- and slime-covered tug was pulled into a relatively upright position and warped into the drydock and set down on the blocks for inspection.

This final phase of the salvage operation took place under floodlights on a rainy night, and the happiest man on the quayside as the tug at last came upright was Sammy Collins.

Rain-soaked, he strode across the wharf, jumped over cables holding the vessel in position and shouted: "She's up; she's up!"

This complicated salvage operation proved to be hugely expensive for Sam Collins – and it also spelt the end of the doughty vessel that the Texan had used in 1961 as a trail-blazer in his sea-diamond mining venture.

The big steam-tug, the workhorse that had served him so well, was never used again and ended up as scrap-metal.

A memento of the *Emerson K* – the tug's emergency steering wheel – could still years later be seen adorning the front wall of the Durbanville home of marine mining engineer Hans Duyzers, who retired in March 1995 after long and distinguished service with Marine Diamond Corporation/De Beers Marine.

Born in the Netherlands in 1933, and trained first as a draughtsman and then as a seagoing engineer, Hans Duyzers headed the engineering department of the marine diamond mining operations off SWA/Namibia for 24 years.

His technical, practical and innovative abilities were, to a large extent, instrumental in the success of De Beers Marine (Debmar), where he played a key role in developing techniques for deep-water sampling operations and mining methods that led to the designing and commissioning of De Beers's current production fleet.

CHAPTER ELEVEN

Life in the floating mining camps, in a cruel sea

THE loss of one barge, the grounding of another and the sudden sinking with the loss of six lives of one of the MDC's tugs, served to highlight the harsh and hazardous conditions in which the seaborne diamond miners often had to work.

In a special feature published by the Afrikaans magazine *Die Huisgenoot* in November 1964, journalist Louw Pretorius wrote:

"Believe me, there are few men who would be able to withstand for long the heavy work-demands and the conditions in these floating mining camps. This really is a life for the hardiest of men."

Pretorius spent a week among Sammy Collins's sea-diamond miners off the coast of SWA, and he described his visit as "an unforgettable experience; something that will remain in the memory for the rest of one's life".

The highlight of this visit, he said, was an overnight stay on board the *Diamantkus*, during which he was shown a double handful of diamonds – about 3 000 in all – which he was told had been sucked up from the seabed in one day.

Pretorius pointed out that the miners at that stage were spending nearly three weeks on the mining units, followed by a week's home leave. "On the mining vessels themselves, shifts last 12 hours and, when circumstances demand it, the mining men may have to work as long as 36 hours at a time, with only short breaks for meals.

"And there is no such thing as being compensated for overtime worked. If there's work to be done, the miners know they are expected to be on the job, and that's that".

As impressed as he was with the onerous and demanding nature of the work on the mining vessels, so was Louw Pretorius impressed by the living conditions on board.

"Where food is concerned, Sam Collins's policy is 'give the men what they want; they work hard'.

"The cuisine provided for these sea-diamond miners is like that to be found in the best hotels in our major cities. Food of the highest quality is provided at every mealtime. Even those working night-shifts get a full "lunch" at 1 am.

"The cabins are small, but neat and comfortable. Recreational facilities are also available – movies, a library and so on, but these are really by the way. The men are here to haul diamonds from the sea. It's no holiday – and they are well rewarded for their labours."

Tom Kilgour recalls his initial "grim nightmare"

What Louw Pretorius experienced in 1964 on board the biggest of the marine mining camps was sheer luxury compared with what confronted Tom Kilgour when, as Collins's technical assistant, he paid his first visit to *Barge 77* off the SWA coast two years earlier.

"We were taken up there from Cape Town in the *Schipa*, and as we neared the end of our trip, very early in the morning, we were all aware of a tremendous commotion. I went up on deck to see what it was all about and there I saw the ugly, rusting hulk of *Barge 77* heaving about in the swells.

"The noise we were hearing was coming from her vibrating screens and generators – and it was deafening. Beyond the barge was the most inhospitable and hostile-looking shoreline imaginable. I was quite horrified; the scene before me was like something out of a really grim nightmare.

"My first feeling was to go back to Cape Town right away – and there were others on board the *Schipa* who felt even more strongly than I did. Some of them refused to get off the vessel, and had to be physically moved into the tiny ferry-boat that had come to fetch us, and in which we were rowed across to the barge, the outboard motor having conked out as it so often did.

"In some cases, men who had arrived as replacements for the crew of the barge had hardly set foot on the heaving platform when they dived off again and started swimming back to the *Schipa*. One can imagine the reaction of members of the barge's crew on learning that their replacements had refused to come on board. This usually meant they had to do another tour of duty before being relieved."

Tom Kilgour says he noted with interest on his visits to the barges that many if not most of the men on board were from inland areas. "A good many of those I saw were from the Free State and the Transvaal, and they seemed to be surviving the harsh conditions better than those from coastal areas – including men who had served at sea before, but in conventional ships, not floating mining camps like these."

"Life's tough on the diamond barges…"

Collins's sea-diamond venture was the subject of a special feature in the now-defunct Johannesburg weekend newspaper *Sunday Express*, in April 1965.

Headed "Life's tough on the diamond barges", the article was written by Neil J Smith, after a visit to the sea-diamond mining area off the West Coast. This is what he wrote:

It's a rough 80-mile voyage down the bleak stretch south of the Skeleton Coast of SWA, from Luderitz to the diamond rigs.

"We should have a fairly calm trip," said Captain Don Foulis of the *Marina*, cocking an expert eye at the wind-ruffled waters of Luderitz Bay.

The *Marina*, a sleek ex-Norwegian ferry boat, shuttles up and down the coast carrying supplies and relief personnel to the barges – the floating diamond mines. She, like the barges, is owned by the Marine Diamond Corporation, the brainchild of Texas millionaire Sam Collins, who is now tapping the untold wealth of the sea-bed off this fiercely inhospitable coast.

The *Marina* slipped away from the jetty, leaving behind the little town with its improbable collection of high gabled German buildings. Luderitz was founded by the Germans some 80 years ago, and has changed little since they left at the close of the First World War.

It still looks like a Baltic fishing village transported by some vindictive giant to this wild desert coast of rock and sand. But Luderitz, for all its arid surroundings, is a paradise compared with the world of the diamond barges.

Five minutes after leaving the shelter of the bay, my camerman and I lay prostrate on our bunks as the *Marina* butted through massive swells, with the wind whipping off cascades of spray and dumping it by the swimming-bathful on the foredeck.

"Always lie on your back..."

What was it someone had told me at the corporation's head office in Cape Town? Always lie on your back. Don't attempt to lie on your side or you'll be sick within minutes. The information was superfluous, as it happened. The curious corkscrew-wobble motion of the *Marina* made any other position but flat on one's back impractical.

Still, there was a certain horrid fascination in watching the sea drowning the sealed porthole, cutting out the light periodically with a filter of translucent green.

A calm voyage?

"Oh, this isn't rough," the first mate told me over a lunch that under any other circumstances would probably have tasted delicious.

"The wind isn't even at gale force. Anyway, our other crew-change ship, the *Schipa*, is much worse. We can almost guarantee you'll get sick on the *Schipa*."

Seven hours of pitching, tossing, vibrating and yawing brought us to diamond territory – or at least the tiny portion of it now being exploited by the corporation.

A Christmas party on one of the Marine Diamond Corporation barges. Mining Superintendent Eddie Redgrave at piano.

In a gentle curve of coastline, half-a-dozen vessels, large and small, bobbed at anchor. The two biggest, namely the *CCC 111* and the *Diamantkus*, are the floating diamond mines. Another barge, the *Colpontoon*, was away in Cape Town undergoing repairs.

She had run aground in February after losing her anchors in a fearsome night of wave mountains and shrieking winds.

Up on the grey beach lay the gaunt skeleton of yet another barge, the "77", which was swept on to the rocks two years ago to become a useless hulk. The mining of sea diamonds is not without its special dangers.

Beside and between the two bigger vessels were the tenders, tugs and tankers that act as the errand boys, nursemaids and delivery vans for the rigs.

But first there was a terrifying leap from an iron ladder down the side of the heavily-rolling *Marina* into a bucking dinghy, with typewriters, cameras and gear stowed away in plastic bags ("We haven't lost a man or a suitcase yet – but you never know!").

And then a breathless plunge through the swells, a leap onto the narrow deck of a tender, and finally a prodigious hop from the tender to the diamond rig, timing it carefully so that one wouldn't land in the icy water, or be crushed between the two bobbing craft.

A floating diamond mine is no thing of beauty. The *CCC 111* (or "One-Eleven") looks like a cross between a river dredger and Noah's Ark. The dredger is the working portion where the gravel, shell and sand of the ocean bed is processed for its rich harvest of gems.

The Ark – a huge domed superstructure – contains the living quarters where 56 men eat, sleep and relax during their stiff 18 days working stint.

The rigs are at present operating about a quarter of a mile offshore, within sight of the breakers and the razor-sharp black rocks that send spiny fingers into the sea from the bleak coast of shifting sand dunes and hard-caked beach.

It is one of the most inhospitable coastlines in the world. Average rainfall is a fraction of an inch. Thick sea-mists often obscure the sun and shore and keep the temperature down to a bracing level.

"Terrifyingly high seas…"

The sea water is a chilly 50 degrees Fahrenheit. Gales are frequent and unpredictable. If they come from the south-west they can bring terrifyingly high seas.

Diamonds have been produced in this area for more than half a century. The theory was that most of the stones had been washed ashore by the sea or lifted from the sea-bed through the ages, as the shoreline rose. But this was only a theory until Sam Collins started his operations four years ago.

Collins, an expert in pipe-laying, first became interested in the possibility of sea diamond mining when his firm tendered to lay an underwater pipeline to supply oil to the De Beers shore workings at the mouth of the Orange River.

He reasoned that if a pipeline could be laid to transport fuel, it would need very little adaptation to turn it into a suction pipeline pumping up the fabulously rich gem harvest that he was sure lay on the sea-bed.

Thus the MDC was born, with the gloomy prophets predicting that this 'harebrained' scheme was doomed to failure in advance.

"Water doesn't stop minerals," Collins riposted to the critics. "If there are diamonds on land then they are lying there on the ocean bed as well."

He was proved right in 1961, when his vessels started bringing up the first gemstones.

The mining process used by the barges is a simple one. Huge suction pipes trailing behind the rigs pump up the sea bed – first the sand and slush and then, as the nozzles of the pipes penetrate deeper, shell, gravel and stone until the pipes reach bed-rock perhaps 40 feet below the surface sand.

The gravel and the crevices in the rock are the richest sources of

Sorters extracting diamonds from the jig concentrate on one of the MDC diamond barges working off the SWA coast.

diamonds. In one memorable strike the *CCC 111* lifted 2 193 carats in 24 hours.

"We were finding 60 to 70 diamonds in each pan," the cheerfully gruff mining superintendent, Bob Greenshields, said wistfully. "It was a real bonanza. Thirty-thousand quids' worth of stones in one day!"

Before the gravel finds its way into the pan it passes through various processes. First, it rides over a vibrating grid that rejects all material more than seven-eighths of an inch in diameter. All the big stuff is dumped overboard.

If another Cullinan diamond were to emerge from the ocean bed off South West Africa, it would return to the ocean depths before anyone could do anything about it.

"But the chances are pretty remote," Greenshields said. "We might find one stone that size in ten years. It isn't worth the extra cost and labour to process everything we pump up."

Next, the small material under one-and-a-half millimetres in diameter is sieved off and discarded. The remaining mixture is pumped into a whirling cone called a cyclone, where a sludgy compound of ferro-silicon and seawater is added. The vortex forces all the heavier material including the diamonds to the outside of the cone, from where it drops gradually to the bottom.

Regularly, at 20-minute intervals, one pan of the heavy material from the cyclone is tipped on to a green baize-covered table in a little wire cage on the main deck of the barge. Two sorters with putty knives sift through the mixture of shell, pebbles and gravel. These are experienced men and can easily distinguish the diamonds from the quartz and other non-precious transparent material.

An ever-alert security man (Colonel Rex Redelinghuys, former head of the Diamond Branch of the CID) picks up the stones with tweezers as the sorters point them out, and chalks up a running daily score.

800 diamonds on an average good day

On an average good day, as many as 800 diamonds may be lifted from the sea bed on *CCC 111*. Most of them are small – the average size is only about half a carat – but they are of a much finer quality than those found ashore.

Perhaps half-a-dozen may be stones of two or more carats, carefully marked down in the tally with a little diamond symbol.

But not all days are good – or even average.

One of the supervisors on the barge explained: "You find the diamonds, the plant is working perfectly, you think you're set for the next few days – then the wind comes up. You are blown around, just enough to move the pipe away from the gems. The wind stops, but you've lost the pay-off position.

"Then you find diamonds again, the weather holds – then the plant breaks down..."

In spite of the difficulties, the corporation has produced almost 400 000 carats of diamonds in its three years of operation. Eventually, it hopes to produce a million carats a year.

Collins estimates the reserves at 40 to 50 million carats, scattered over his 14 000 square miles of ocean-bed concession. So far, three years of intensive mining operations have covered just 150 acres.

"Inexhaustible is a pretty strong word," he says, "but I reckon the corporation will be mining diamonds off this coastline long after I'm not around any more. There are enough diamonds down there to last a hell of a long time."

Security is a problem on the rigs, though not as great as it would be ashore, where X-ray checks and fierce police dogs help protect De Beers against theft.

On the barges, the precautions seem almost elementary. The sorting house is merely a wire cage. It is not locked, but it is never empty, and there are generally three men in it at a time.

"Two men might get into collusion with each other, but with three the chances are far less," is the disarmingly frank explanation.

One of the men in the sorting house is always a security officer, keeping a watchful eye on the sorters. But security on the rigs does not end with the man with the tweezers. Security duties are also assigned to other crew members, and in Cape Town the security staff keeps careful check on the main illicit diamond buyers.

"Anyone who wanted to get away with stones would have to sell

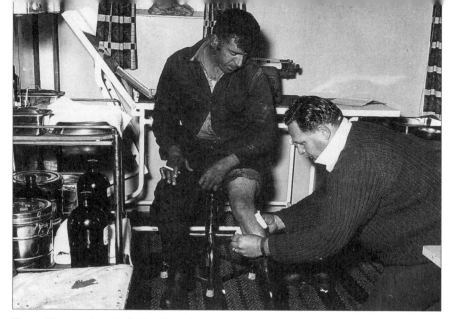

First-aiders on board the hard-worked diamond barges were kept busy tending cuts, abrasions and other injuries.

them in Cape Town," Collins told me. "We usually hear about, and crack plots of this kind before they get off the ground."

Yet, surprisingly enough, on the rigs it is often hard to remember that sparkling gems are the reason for the whole complicated set-up. Only about seven of the 56 men on the *CCC 111* have any direct physical contact with the diamonds. The others do a variety of jobs ranging from welding to diving, and from electrical maintenance to laundering.

It could almost be a cramped, grubby and noisy seaside workshop with living quarters attached.

No women allowed on a working diamond rig

The world of the diamond rigs is a tough, enclosed, masculine world. No women have ever been aboard the barges while they are working. In fact, the thought of high heels clicking across the oil-slicked decks is almost too grotesque to consider seriously.

The rigs are small, and cramped. Every inch of space is used – for heavy equipment, conveyers, winches, storage. There is little room to spare for comfort, and certainly none for luxury.

The men work to a cycle of 27 days: 18 days of long 12-hour shifts, followed by nine days shore-leave. The corporation takes the outgoing crew to Luderitz aboard the *Marina* or the *Schipa* and then flies them to Cape Town in its own Dakotas.

Life falls into a fixed routine aboard a diamond rig. Always in the background is the hum, roar and clatter of machinery. The days are punctuated by the coming and going of the tugs and tenders, the rise and fall of the sea, the slap of the swells against the flat bottom of the barge – and the wind.

The wind seemingly never stops blowing on this bare, grey coast. Even in the tranquil times of the year it is unpredictable, liable to reach howling gale force within a few hectic hours. In winter, the damp chilly gales press hard on each other.

The sun always shines on the desert shore, except when the dense sea mists roll in. But always, there is the wind, and the heaving of the well-anchored barge in the rolling swells.

Life is as comfortable as it may be in cramped quarters. Food is plentiful and good. After work there are nightly films, letters to write, books to read, chess, darts and card games to play – or perhaps a little fishing from the deck. But a 12-hour working day doesn't leave much time for idleness or for building up tensions among the crew.

"We seldom have fights among the boys," Bob Greenshields said. "They work too hard, and anyway 18 days isn't all that long. It was different in the Navy during the war, when you might be cooped up with the same faces for two years or more."

Why men opt for this spartan way of life

What induces a man to give up a comfortable life ashore for the spartan existence of the diamond rigs? This is life at sea without any of the compensations of exotic landfalls and exciting ports. No blue cloud-capped mountain peaks and waving palm trees, no night clubs and dockside dives – only the steeply shelving sand and grim black-and-grey landscape of the Namib Desert, a few hundred yards away.

The pay certainly has something to do with it. Many of the men on board the *CCC 111* earn at least a third more than an untrained man could expect to earn ashore in South Africa.

And those actually engaged in mining operations have the added incentive of a bonus paid whenever production rises above a set quota.

But basically, the reason for the presence of these men on board the barges is man's hankering after adventure. As long as there are young men, and men not so young, whose eyes are fixed on that far horizon in a world of ever-diminishing horizons, people will come to work on the diamond barges.

As long as there are people whose personalities cry out for the freedom – and discipline – of living the hard life of a man among men, the *CCC 111*, the *Colpontoon* and the *Diamantkus* will not want for crew.

As long as an aura of glamour attaches itself to the very word "diamond", men young and old will sacrifice the easier way of life ashore for the world of the diamond rigs.

CHAPTER TWELVE

Vema Seamount, Tretchikoff – and HF Verwoerd

AS Francois Hoffman has put it, Sammy Collins was an "easy touch" for people with ideas – and challenges.

And when it was suggested to him that the Vema Seamount in the South Atlantic, 400 nautical miles north-west of Cape Town, could be a rich source of diamonds, he decided he had to go there and investigate, using the *Emerson K*.

This was in November 1964, just before a series of setbacks and disasters brought about a change of fortune for Collins and Marine Diamonds.

Sammy also decided to put the tug at the disposal, on the same trip, of scientists wanting to investigate commercial fishing prospects in the Vema area, already known to be rich in marine life.

Gary Haselau recalls that Collins had the *Emerson K* fuelled, victualled and fully equipped for a week's scientific prospecting on this underwater mountain peak off the Southern African coast, which rises about 5 300 metres above the floor of the Atlantic and is only about 28 metres below the sea-surface.

"In that week we did a lot of diving and probing among the peaks of this massive mountain range under the sea," he says. "And although we saw masses of crayfish, we found no diamonds on Vema or any of the other peaks."

In honour of Sammy's visit to Vema, one of the undersea rock pinnacles discovered was named Collins Peak. And while the Texan may have failed to find the diamonds he had set out to discover, his trip was regarded as a resounding success in other respects.

It was reported at the time that the *Emerson K*'s visit to Vema Seamount was one of the most interesting scientific-commercial projects to be undertaken from Cape Town in recent years.

The expedition was under the auspices of the newly-formed South African National Committee for Oceanographic Research, and was led by Professor ESW Simpson of the Department of Oceanography at the University of Cape Town.

It found definite evidence of the volcanic origin of the Vema peak, which juts up from a wide plateau – later named the "Emerson K Plateau" in honour of the tug's visit – which has deep gullies.

Divers went down from the tug and collected evidence of fish life in the area, and came up with some interesting facts. Among these was that abundant fish-life on the peak is related to that found around Tristan da Cunha, and not to that found along the Southern African coasts.

On the tug's return to Cape Town Sam Collins said in an interview that, as the Vema Seamount – so called because it was discovered by the US research vessel *Vema* – was outside the influence of the Benguela Current, it presented a totally different picture from that in South African waters.

"Maybe," he said, "this is the long-lost mother-source of sea diamonds. We don't know yet, but there is a possibility that it is".

Tretchikoff

Sam Collins became an avid fan of the Russian-born artist Vladimir Tretchikoff, whose paintings immediately took his fancy when he first saw them – first in Paris, then in Cape Town, the city of Tretchi's adoption.

Original works by Tretchikoff adorned the walls of Sammy's offices and his homes, in Cape Town and London.

One of those that hung in his office on the Cape Town Foreshore, "African Drum", depicts a tribesman beating a drum against a flaming-red background. It was said he paid R8 000 for the original – and a probable reason he liked this painting so much was that it suggested action, speed and movement – the things Sammy liked in others and valued in himself.

Collins, Tretchi and their wives became close friends, and the wealthy and controversial artist ("I laugh all the way to the bank!") also became a substantial shareholder in Marine Diamonds. Tretchi recalls how they met:

"Around about 1960, Sammy and Gigi were in Paris and they saw a framed print of my 'Lost Orchid' in the window of an art shop. They immediately fell in love with it, and when Sammy inquired inside where he could get the original, he was told this would not be possible, as 'the artist, Tretchikoff, is dead and we don't know where the original is'.

"This art dealer had probably read or heard about the near-fatal accident I was involved in, when my Cadillac convertible rolled on a Cape Town highway and I was pinned, unconscious, underneath.

"They had quite a job getting me out of the wrecked car, and I was at death's door for quite a while. I made what was said to be a miraculous recovery, and sometime later when I was up and about again a newspaper reporter told me he had written my obituary, as everyone thought I was a gonner for sure.

"That Cadillac convertible of mine, incidentally, had been bor-

Sammy and Gigi (right), with Vladimir Tretchikoff between them, show sea diamonds to the cast of My Fair Lady.

rowed for use in the VIP motorcades when British Prime Minister Harold Macmillan and his wife Lady Dorothy visited the Cape in 1960.

"Later, when Sammy Collins came to Cape Town, he heard there was a Russian artist called Tretchikoff living here. He immediately thought of the framed print he had seen in Paris, and wondered if this could be the same Tretchikoff – the one he had been told was dead.

"He looked me up in the telephone directory, and was delighted to discover that Tretchikoff – the same one who had painted 'Lost Orchid' – was indeed alive and still painting. He and his business partner Emerson Kailey came to see me at my home. We got on very well together, and this was how a close friendship and business relationship developed between Sammy and myself.

"I remember that at an early stage, when Sammy came to my home and we got round to discussing his idea of mining diamonds from the sea, we took some cocktail peanuts out of a bowl on the coffee table to demonstrate our respective ideas of how the diamonds would be lying on the seabed.

"I was so impressed by Sammy's sea-diamond plans that I offered at that early stage to put £50 000 into the venture. I became a shareholder in Marine Diamonds, and I've still to this day got shares in their successor, De Beers.

"I wasn't able to help Sammy buy the original of 'Lost Orchid' because it had already been sold in New York, and its present owner was the Johannesburg millionaire John Schlesinger who had bought it for 125 guineas.

"In any event Sammy eventually bought quite a few of my paintings; I think he ended up with about 16 originals.

"He saw eight of these paintings in my studio at home, and wanted to buy them there and then. When I told him they had already been catalogued for an exhibition at Harrods in London and were about to be sent over there, he arranged to have them reserved for him in advance, and he ended up owning them. Emerson Kailey also bought some of my paintings, for his own home.

"That exhibition at Harrods broke all the existing attendance records at the time. An average of 7 000 people a day saw my paintings during the show, and by the time it closed, the total attendance figure had risen to 205 000."

A memento that Tretchi has of his close friendship with Sammy Collins is the two Persian carpets that adorn the lounge floor of the mountainside home at Bishopscourt that he bought at the height of his fame in 1967.

These had been presented to Collins by Saudi-Arabian oil-rich sheiks in appreciation of a submarine oil pipeline he had laid for them in the Persian Gulf.

The "floating gin-palace"

The handsome little *Schipa*, when she was not being used as a crew-change ferry in the sea-diamond venture, virtually became Sammy Collins's private motor yacht, and as such she left a wealth of raunchy stories and anecdotes in her wake.

She came to be known as "Sammy's floating gin-palace".

Schipa had been built during World War II as a Fairmile river and coastal patrol craft, and Collins's business associate Bill Mitchell had had her converted into a deep-sea pleasure cruiser, equipped for big-game fishing.

Her original engines had been the type used in wartime armoured cars. Mitchell replaced these with Rolls-Royce engines, which turned out to be unsuited to this type of vessel, causing recurrent back-pressure problems. They were replaced by GM diesels.

After its conversion the vessel looked in profile like a miniature ocean liner. It became noted for lavish parties, and for the number of pretty women and celebrities who were hosted on board. But *Schipa*'s light, wooden construction made her lively in anything but calm conditions, and she was also noted for the number of passengers who became horribly seasick on board.

At one stage her captain was George Foulis's brother Don. His chief officer, Eddie Redgrave, was a veteran of 24 years service with the Royal Navy, and before joining Marine Diamonds he had been a navigating officer with Thesen's Steamship Company.

Eddie, who retired to his home at Bakoven, Cape, after working

Sammy's wife Gigi makes a presentation to Prime Minister Hendrik Verwoerd, at a Parliamentary Angling Club lunch.

for Sam Collins and finishing up as a mining superintendent, recalls that Sammy insisted on his officers wearing proper nautical uniform, replete with rank-insignia.

"We had to be at Sammy's disposal whenever he required *Schipa* for his or the company's use, and we had some very interesting people on board as passengers.

"I remember that on one occasion he ordered us to head for Saldanha Bay, and then to look for a nice sheltered spot there at which to anchor, as we had a special guest joining us for the weekend.

"This happened to be the weekend in November 1963 that President Kennedy was assassinated in Dallas, Texas, and when he heard the news over the radio, Sammy was terribly upset. He was crying, in fact, because not only had his country lost its President but this terrible thing had happened in his home state, not far from where he grew up.

"He seemed to have known Jackie Kennedy, and he had a message of condolence to her radioed ashore from the *Schipa* at Saldanha, for relaying to the United States.

"On another occasion Sammy and his wife 'Gigi' entertained the entire cast of the play *My Fair Lady* on board – and what a spread there was for them!

"Sammy used *Schipa* a lot, and quite often he would get us to anchor the vessel off Robben Island when he had special guests on board. I remember that Tretchikoff and his wife were on board on at least one such occasion, and a bevy of pretty girls would sometimes be included on the guest-list."

Sammy's admiration for Verwoerd

Gary Haselau, as Sam Collins's PRO, at one time had the job of taking VIPs tunny fishing in *Schipa*.

"MPs belonging to the Parliamentary Angling Club were among the celebrities I used to take out on such occasions. The then Prime

Minister, Dr HF Verwoerd, happened to be a member of the club. (Another prominent member at the time was John Vorster, Minister of Justice, who succeeded to the premiership after Verwoerd's assassination in 1966. He too was taken to sea on occasion in the *Schipa*).

"Sammy thought it would be a good idea to present the club with a trophy for tunny-fishing, so he promptly had one made, and presented it to Verwoerd. This was a very handsome, hand-crafted cup surmounted by a leaping tunny. Verwoerd was obviously impressed – and delighted."

Collins, a true son of the Deep South, not only admired Dr Verwoerd as an angler; he also espoused the political philosophy and segregationist policies of the Prime Minister and his ruling National Party.

Marine Diamonds director Peter Keeble also became actively involved in National Party politics.

Collins said he had visited and worked in many countries that had weak governments. South Africa, by contrast, had a "strong" government, and for this reason it had his full support.

He made substantial contributions to National Party funds while he was in South Africa, and his motive for doing so was questioned in Parliament by the opposition United Party.

When Dr Verwoerd said in a speech in the House of Assembly that South Africa's prosperity was being realised by visitors from abroad "such as Mr Sam Collins", a UP front-bencher, Marais Steyn, interjected: "Yes – people who receive diamond concessions!"

Steyn was reprimanded for this insinuation by the National Party member for Malmesbury, JW van Staden, who said Steyn's remark had been "low", because Collins was not able to defend his name. He added that Collins had not received a diamond concession from the government but had in fact bought it through a group that had included Abe Bloomberg, for some years a United Party MP.

Collins sparked a polemic in the press in 1964 when the *Cape Times* published a letter in which he expressed support for the government's policy of *apartheid*. "I believe the Nationalist idea of separate development is a step in the right direction," he wrote.

Expounding his views on legally enforced separation of the races, he said a more rigorous form of *apartheid* than that to be found in South Africa was practised in India, and that "I can state categorically that in no nation in the world have I seen as happy and contented a lot of the lower-class people as I have seen in South Africa."

Collins's right-wing views brought an immediate reaction and were strongly challenged and denounced by a number of newspaper readers. These included J Moresby-White of Sir Lowry's Pass in the Cape, who pointed out that the Indian form of discrimination was based on convention and on caste, not colour.

Sammy presents a trophy for deep-sea angling to Herman Martins, MP, of the Parliamentary Angling Club. (Mrs Betsy Verwoerd in foreground).

"Though Mr Collins appears to have misread the political situation here," he wrote, "he and his wife are indeed wise to make South Africa their home, because in spite of the many trials and tribulations that lie ahead of us because of National Party policy, South Africa can and undoubtedly will find before too late the road to real peace and prosperity for all her inhabitants – white, brown and black."

Prophetic words, back in 1964?

CHAPTER THIRTEEN

The great sea diamond rush gains momentum

IN proving that seabed diamonds could be mined, whether economically or not, Sam Collins unwittingly triggered a great sea-diamond rush, with one financial group after another wanting to get in on the act.

In mid-1963 the South West Africa administration announced that sea diamond concessions for roughly 1 000 km of the SWA coast had been granted to a new company called Terra Marina.

This new venture was backed by Afrikaner finance, and it now meant that the entire SWA coastline was covered by offshore concessions from the low-water mark to the three-mile limit. The complete SWA offshore situation was now:

1. From the Orange River to Luderitz – Sam Collins.
2. Luderitz (Diaz Point) to Hottentot Bay – Terra Marina.
3. Hottentot Bay to south of Sandwich Bay – ME Kahan, backed by US oil tycoon J Paul Getty.
4. Sandwich Bay to the Cunene – Terra Marina.

The Terra Marina consortium comprised Bonuskor, Federale Mynbou, Duineveld Beleggingskorporasie (which controlled Weskus Mynbou), Spes Bona Mining Company and Federale Volksbeleggings.

Grants were made to it by the Executive Committee of the SWA Administration after it had heard a deputation comprising Dr MS Louw, a prominent South African financier; Dr Piet Neethling and Mr AP du Preez of the Duineveld group; Mr PH Meyer, MPC for Bellville, representing the Spes Bona interests; and Mr BD Maree, chief geologist of Federale Mynbou.

It was reported at the time that the new allocations to Terra Marina were "particularly valuable when one considers the rich diamond finds near Plum Pudding Island by Sam Collins.

"The spheres of influence of Collins and Terra Marina will no doubt overlap, and special arrangements will have to be signed if they want to avert a clash."

Apart from Plum Pudding, the most important South Africanadministered islands falling under Terra Marina's new concessions were Halifax, North Long, South Long, Possession, Pomona, Sinclair and Roast Beef.

The concessions awarded, from Diaz Point to Hottentot Bay, and from the northern boundary of Diamond Area No 2 right to the Cunene River – SWA's border with Angola – were north of, and several times larger than, the concession already being exploited by Sam Collins's Marine Diamond and Sea Diamond corporations.

On the technical side, it was reported that Terra Marina was ordering a dredger and modern equipment from the Netherlands, at a cost of R900 000. Meanwhile it would use a converted pilchard fishing boat, *Ontdekker I*, which was being equipped at Hout Bay to probe the seabed to a depth of 60 metres.

Terra Marina was said to be departing from MDC technique by divorcing dredging from sorting and recovery. Once diamond-bearing gravel had been scooped up by the dredger and screened, it would be taken to land-based processing plants – probably at Baker Bay and also at Luderitz.

A problem foreseen for Terra Marina was that its islands were scattered along a vast stretch of coast. If its operations extended much beyond Plum Pudding and Sinclair (both near Baker Bay) the gravel would have to be transported long distances.

The swift move-in of the Terra Marina groups on the diamond front was reported at the time to have left Collins "a little surprised," and "somewhat ruffled".

He was quoted as saying that both he and De Beers had applied for the stretch from Luderitz to Hottentot Bay but had been told that the SWA Administration would withhold this section from claimants until 1966.

The common factor in both major sea-diamond enterprises was the Strydom/Bloomberg/Du Preez-Neethling group. With a 12,5 percent share plus a 5% production royalty in Collins's Marine Diamond Corporation, and a large share in Terra Marina via Duineveld, Strydom, Neethling, Du Preez and Bloomberg had got their group neatly entrenched in the centre of sea-diamond activity.

They now held a balance of power which could pay significant dividends in the future.

A "hell of a fight" over Eiland Diamante

Neethling and Du Preez had discovered that there were certain islands off the SWA coast that were not part of the mandated territory but which, with Walvis Bay, belonged to South Africa. They learnt moreover that these islands had their own territorial waters which were excluded from Collins's Marine Diamonds concession.

Once again, they decided to go for the gap. They applied to the

South African government for a separate "islands" concession and got it, in the name of a new company called Eiland Diamante, which became a subsidiary of Terra Marina.

Leading Afrikaner financial institutions including Sanlam, Trust Bank and Federale Volksbeleggings were induced to take shares in both companies.

Francois Hoffman recalls that "there was a hell of a fight over this because, however legitimate these formal agreements may have been, there had been a tacit understanding between the two groups that they were into these things together.

"Yet here were Du Preez/Neethling getting formal approval to mine sea-diamonds right inside the Collins concession.

"This became a really big bone of contention at the time, with Collins demanding to know why the Du Preez/Neethling group had done this thing without first offering it to him, or even asking him to join them in their application.

"In any case, Eiland Diamante went ahead with their plans to mine diamonds around the islands – but they also lost their boots in the effort. They built a jetty at Baker Bay, but the sea in this area is very unpredictable, and in one of its more violent moods it took the jetty away. A plant that the company built on land was also lost."

Dr Piet Neethling recalls that for a few months, the Eiland Diamante operation recovered a lot of diamonds of good quality, around the islands. "And Anglo American's CDM were watching us all the time with hawks' eyes. They watched precisely what we were doing, and where we were doing it.

Black Sophie's diamond "necklace"

"During our operations we came across a rocky projection from the sea by the name of Black Sophie, and we found a lot of diamonds there. We were taking out something like 2 000 carats a month there, for several months – all of them of good quality, and between one and two carats.

"Now we found ourselves faced with the problem of whether Black Sophie fell within the definition of an island, and there was even a move to take us to the World Court on this, because we could not prove that this was in fact an island.

"One of the definitions of an island is that one part of it must *always* be dry, and this being so it was now up to those wanting to take us to court to prove that Black Sophie was at times completely covered by the sea.

"In any event we just carried on mining there, as well as around islands such as Pomona and Plum Pudding.

"Then, the strangest thing happened. We had a system whereby our mining people worked for three weeks, then took one week off.

Just when we thought we were doing well, with good yields of diamonds from the seabed, it was time for our people to go on shore leave.

"We had been using shore-based electronic apparatus to determine exactly where we were working at any time – yet when our workers returned, to resume operations, they could not find the exact spot again, where so many diamonds had been detected.

"The result of this was that the whole operation went right down. We landed in financial difficulties, and we had to dispose of Eiland Diamante and its associated companies.

"A lot of people lost money, but we had made it clear from the outset that this was a speculative venture.

"Quite a substantial sum of money was owed by our holding company, Duineveld Beleggings, to Caltex for fuel supplied, and AP du Preez and I made a point of paying this off in our personal capacities, as we could not leave things like that.

"The companies were eventually sold to what is now the Trans Hex Group, of which Francois Hoffman was till recently the chairman."

The American oil tycoon J Paul Getty, said at the time to be the world's richest man, became the latest international financier to be attracted by the prodigious diamond potential in the sea off the South West African coast.

In Johannesburg, the Tidewater Oil Company of America – a wholly-owned subsidiary of the Getty Oil Company – had signed an agreement under which it was granted the right from the Industrial Diamonds Group to mine for diamonds in certain concession areas of the Diamond Mining and Utility Company Ltd. The chairman of both these companies was ME Kahan.

It was said at the time that the Getty interests paid about R2,8 million to get control.

Essentially the American group would be given a share interest in the holding company, and pay the group royalty on all diamonds recovered. This was believed to be 50 percent more than the 12,5% previously negotiated by the Industrial Diamonds Group with Sam Collins, but which was not put into effect.

The rights taken over by the Tidewater Oil Company included concessions to mine for diamonds on land as well as in the sea, for a distance of six miles offshore.

Tidewater Oil registered a wholly-owned subsidiary in South Africa under the name Veedol Minerals (Pty) Ltd, to exploit the concessions taken over by them from the Diamond Mining and Utility Company.

Ocean Science – and 'Rockeater'

Meanwhile De Beers, in a project aimed at evaluating the offshore concessions as fairly and quickly as possible, had engaged an American firm called Ocean Science and Engineering to carry out a systematic sampling and mapping programme.

A small ex-US Army coaster was acquired and fitted out as an exploration and sampling vessel. Renamed *Rockeater*, it started operations in 1964, under the command of Captain CJ Harris of Cape Town. A mapping programme had been started late in 1963 and eventually covered the area between Olifants River and Conception Bay, just south of Walvis Bay.

Diamonds are notoriously elusive, and acres of worthless gravel can lie right alongside a rich find. *Rockeater*'s task was to narrow the search to the most-likely parts of the concession areas to yield diamonds in payable quantities.

The plant the vessel carried was capable of screening and sorting gravel at the rate of 20 tons an hour.

In the geophysical method of mapping, recordings are made of sonic echoes reflected from the interfaces of horizons of differing density. The sound is generated by some source of energy such as the discharge of a spark, the thump of a piston or a pulsing diaphragm.

The echoes are "heard" by an underwater microphone (hydrophone) and processed by a recorder which draws a profile of the seabed on a strip of paper. These profiles are interpreted and the results plotted to yield maps that show the water-depth, unconsolidated sediment thickness and elevation (below sea level) of solid bedrock.

Rockeater was thus able to sample the most favourable area of potential diamond accumulation, and could help in avoiding areas where the sediment was too thick, or devoid of sediment cover.

The research vessel sampled by means of drilling. A drill-bit fitted with spikes and water-jets for loosening the material was lowered at the end of a 15-cm (6-inch) drill pipe, suspended from a tower positioned over a hole in the middle of the ship. Sand and gravel were sucked up through the drill-pipe and discharged onto screens, to get rid of the boulders and fine sand.

The gravel was mixed with ferrosilicon – which resembles iron filings – and water, and the mixture pumped through equipment which separates the lighter material such as shell and quartz from the heavier grains such as diamonds and iron-rich rocks. The heavier material was finally hand-sorted and the individual diamonds recovered.

Rockeater sampled along lines oriented at right-angles to the coast. The vessel was equipped with anchors which could be dropped in a pattern determined by the prevailing wind and sea conditions and then winch herself along the sample line, drilling holes at fixed intervals.

Rockeater, *the vessel used in scientific survey work off SWA. She was commanded by Captain C J Harris, of Fish Hoek.*

Assisting *Rockeater* in its search for diamonds was a small converted fishing trawler, the *Klipbok*, which surveyed the ocean bed before sampling. This vessel drew along the ocean's surface a highly sensitive fathometer-type instrument housed in a robust casing. This instrument used a sound pulse to record on a graph the outline of the bedrock and also the overlying sediment.

This information was passed to *Rockeater* to help in formulating the drilling programme.

Klipbok also carried a closed-circuit underwater television camera, which gave a picture of the ocean bed over which the ship was operating.

CHAPTER FOURTEEN

De Beers takes over

Let us now praise famous men – men of little showing. For their work continueth; broad and deep continueth. Greater than their knowing!

— Stalky and Co. – school song

IN May 1965 De Beers was able to exercise its option and purchase a majority share in Collins's Marine Diamond Corporation.

De Beers announced that Consolidated Diamond Mines would transfer to Marine Diamonds its concession over the coastal strip in SWA and pay R495 000 in exchange for 29 percent of the Marine Diamonds issued share capital.

In addition, De Beers made an interest-free loan of R6,5 million to Marine Diamonds to finance the estimated capital expenditure required for the sea area and for the coastal strip, and Anglo American agreed to become consulting engineers for the projects.

By subsequent agreements De Beers, through CDM, acquired further direct and indirect interests in Marine Diamonds, making Collins's corporation a subsidiary of CDM. These agreements effectively shifted control of the Marine Diamond Corporation from Cape Town to Johannesburg.

The take-over happened at a time when the De Beers group was doing all it could to raise production. The value of diamonds available for sale through the group's Central Selling Organisation had risen by nearly 40 percent between 1962 and the end of 1964, and it was anticipated that there would be a further increase in the next two years.

This growth in value had caused a shortage of diamonds, and at the time of the take-over no reserve stocks were held. Harry Oppenheimer said in his annual De Beers statement in 1965 that world demand for gem diamonds was likely to expand further.

After the De Beers deal with Marine Diamonds, the MDC's capital was split up this way: Consolidated Diamond Mines 29 percent, Sea Diamond Corporation 29%, General Mining 13,35%, Diamond Royalties 13% (plus royalty rights), Anglovaal 8,54%, Middle Wits 4,27% and minor shareholders 2,83%.

Collins retained 3 421 shares in Sea Diamond Corporation and it was agreed he would continue as chairman until November 30,

De Beers director Ted Brown (right), Abe Bloomberg (left) and Sam Collins when De Beers took MDC over in 1965.

1966. He would remain a director after that. He would also remain a director of Marine Diamonds but had resigned as managing director. The boards of both companies were to be reconstituted.

As a result of certain share issues, CDM acquired effective control not only of Marine Diamond Corporation but also the following companies: Orama Holdings Ltd; Panther Head Investments (Pty) Ltd; Chameis Bay Holdings (Pty) Ltd; Atlantiese Diamant Korporasie Bpk; Suidwes-Afrika Prospekteerders (Edms) Bpk and Sea Diamond Corporation Ltd.

Dr Piet Neethling recalls that Collins and other directors of Marine Diamonds went to Johannesburg from Cape Town on numerous occasions before and during the take-over period for talks with the heads of Anglo American.

"Harry Oppenheimer himself attended these meetings, and his associates Julian Ogilvie Thompson and Ted Brown were usually there as well."

Cape Town chartered accountant Danny Ipp, Collins's financial chief as well as a director of Sea Diamond Corporation, also played an important role in the take-over negotiations with De Beers, before he left the MDC to return to his accountancy practice in Cape Town.

Ipp had throughout the Marine Diamonds operation kept a meticulously detailed record of all diamonds recovered from the sea, and of all the corporation's operations and transactions. He had exerted a steadying influence on an operation that could have ended in complete chaos financially.

The MDC's Cape Town office was moved from the Heerengracht to new quarters in a building off the Hertzog Boulevard, adjoining Olivetti House on the Foreshore, and accountant Sidney Kagan recalled that "while life may have become less exciting for us, it at least became a lot easier, particularly where our working hours were concerned.

"Where under Sammy Collins we never had fixed working hours, now we worked normal nine-to-five days, with no overtime or weekend work. A result was that my rose-garden suddenly started looking better than ever!

"Bunny Milner became our office manager, and we settled down to a more regularised routine than we had known when we were in the Barclays Bank building on the Heerengracht."

At the time of the take-over the mining vessels operating in Collins's concession area were *Colpontoon*, *Diamantkus* and *Barge 111*. These units were all dredging with freely suspended airlift pipes of various sizes, and employing the inefficient "search/mine" principle.

'Pomona' – the biggest and most complex barge yet

Mining techniques had meanwhile been considerably improved, and De Beers immediately started designing and developing a new and technologically updated mining unit, the *Pomona*, which was commissioned in June 1967 at a total cost of just over R3 million.

At the time, *Pomona* was reported to be "the biggest, heaviest and most complex vessel yet built in South Africa". The design of this massive, topheavy-looking unit was based on the extensive experience gained from *Diamantkus* and Collins's other mining vessels, and it incorporated a new type of diamond recovery boom, described as a breakthrough in offshore recovery techniques.

Two 4 400 hp Mirrlees engines from the *Diamantkus*, still in perfect condition, were installed aboard *Pomona* for power generation, coupled to Brush generators to give an output of 3 020 kw each.

The steel hull of *Pomona*, which displaced 4 800 tons fully loaded, had an overall length of 86,8 metres (285 ft), a beam of 18,2 m and a draught of 4,8 m. The vessel was built in two sections which were floated together in the Sturrock graving dock in Cape Town docks, and the hull was subdivided by massive lateral and longitudinal steel bulkheads.

Although not a fully self-propelled vessel, *Pomona* had an emergency propulsion system comprising two conventional propellers, each driven by a 700-hp AEI electric motor. Both the anchor winches and the propulsion plant were controlled from a central station.

The crew accommodation included 120 bunks in separate two-

The giant R3 million mining rig Pomona, *commissioned in 1967 by De Beers after it had taken over Marine Diamonds.*

and four-berth cubicles, equipped with forced ventilation and arranged in a five-deck house at the forward end of the barge. One complete deck was allocated for recreational purposes and the structure was topped by a landing deck for helicopters.

More-efficient equipment had been developed by De Beers for lifting the seabed gravels onto the ocean-going recovery vessels, and it was anticipated that when this had been installed, output should within four to five months increase to about 35 000 carats a month, and ultimately to between 40 000 and 50 000 carats monthly.

Despite the poor condition and relative inefficiency of the two barges *Colpontoon* and *Barge 111*, it was necessary to keep them in production until the commissioning of *Pomona*.

Also included in the take-over by De Beers was the scientific prospecting vessel *Rockeater*, which went to work sampling areas of interest in greater detail.

Apart from Chameis Bay, mining operations were also undertaken near Plum Pudding Island and in Hottentot Bay just north of Luderitz.

Heavy financial commitments

For the Marine Diamond Corporation, increasingly costly sea-mining operations had not been as successful as anticipated, and rough seas had continued to hamper production. The corporation finished up with mounting cash-flow problems.

In January 1964, 16 months before the De Beers take-over, the *Financial Mail* had reported that diamond production from the MDC's *Barge 111* and the bigger *Diamantkus* had "hit new peaks".

"Last weekend the combined production on one day was 1 397

carats – and MDC experts estimate that the whole organisation needs only 500 carats a day to break even.

"The *Diamantkus* (target production 2 500 carats a day) is, after initial teething troubles, edging its daily figure up towards 1 000. It is still having production problems and only two or three of the six airlifts have been used in the past week.

"But its total of 3 100 carats for the first two weeks of production has been eclipsed by the figures for the past week, which put the *Diamantkus* output at close to 4 000 carats in under a week and *Barge 111*, which is sucking up far more diamonds than had been expected, at nearly 2 500 carats. The total for the six days is over 6 000."

Offsetting this apparently good news was a warning sounded by the same journal, and at the same time, of the MDC's "heavy financial commitments in its sea-diamond mining operations".

In a special "Offshore Diamonds" supplement it was pointed out that capital tied up in the company's fleet alone was over R7 million, and that altogether, more than R10 million was involved in its operations. At that stage Marine Diamonds owned or leased five floating mines, five tugs, four auxiliary vessels for transporting supplies and crew, and two aircraft.

More than 500 people were on the MDC payroll, and 1 500 indirectly employed through business given to other companies.

"Marine Diamonds' all-in costs, including a 10 percent allowance for depreciation and contingencies, are running at R300 000 a month," the *FM* disclosed. "The *Diamantkus* alone costs R4 000 a day to operate and a single prospecting vessel can cost R800 a day.

"As the colourful Mr Collins puts it: 'One boat, two strings of racehorses or three mistresses – they're all damned expensive!'

"At this stage the investor who puts his money into Marine Diamonds (via one of the holding interests) is accepting a big risk, and if he backs one of the concession groups he is risking the loss of every cent against the chance of a handsome killing.

"The most reassuring fact about offshore diamonds is that two big groups – General Mining and Anglovaal – think they are worth risking money on."

The uneconomic *Diamantkus*, incorporating major design faults that were not foreseen during its conversion, was withdrawn from service in June 1966 because of unsafe hull conditions, and was later sunk in a military exercise off Robben Island.

Several former Marine Diamond Corporation executives have cited the high cost of operating and maintaining the former US tank landing craft as the main cause of MDC's decline.

"The logistics of running a venture like this became formidable," says Francois Hoffman. "Our operating costs were enormous, and *Diamantkus* was responsible for about 80 percent of these costs. This is what finally killed the whole venture. Meanwhile Collins had taken his share in Marine Diamonds and put it in Sea Diamonds Limited, and had this listed on the London and Johannesburg stock exchanges.

"In the end he had to go to De Beers which, by then, had had to revise its initial viewpoint on the mining of sea diamonds. We had by then been able to demonstrate, amply, that there were in fact diamonds in recoverable and payable quantities on the seabed.

"So, De Beers came to Collins's rescue, at the end of one of the most remarkable episodes in the history of mining. Sammy Collins, the pioneer of sea-diamond mining, had gone from nothing to boom, and from boom to nothing in the short space of four years.

"Yet the venture that he started – the large-scale recovery of diamonds from the seabed – continues successfully to this day."

Back to pipelines – and into other ventures

While claiming he was "still the boss" in MDC – he saw the De Beers deal as giving him a foothold in SWA – Sam Collins now effectively withdrew from hands-on management of the venture he had launched in 1961.

However, the entrepreneurial spirit still burned strongly in the tough little Texan and apart from going back to his speciality – the construction and pulling of undersea pipelines – Sammy's name became linked with a number of diverse ventures.

In 1965 it appeared in connection with a R700 000 Reef Nigel project to recover diamonds from the bed of the Vaal River, and it came up again with an announcement of the possibility of a major oil find in Zululand, by the Argus Oil Exploration Company of which Collins was chairman.

About the same time, Sammy was also reported to have had talks with Unie Vleis, with a view to entering the deep-sea fishing industry.

Bill Webb says that although Collins was not actively involved in the construction and laying of the oil pipeline from Durban to the Reef, "it was Sammy's idea that such a pipeline be built – and the government acted on his advice."

In May 1965 an observer of the business and financial scene wrote:

"Mr Collins might be forgiven for a typical Texan ability to 'think big', but I think that even his most ardent admirers must admit that the substance of his empire, at this stage, is smaller than the emperor's painting of it.

"On realistic terms, the substance of the empire to date can be boiled down to his 29 percent interest in Marine Diamond Corporation, via his 78% interest in Sea Diamonds.

"On the question of who has what say in Marine Diamonds it would appear that Mr Collins – despite his position in the reconstituted MDC – will not have it all his own way.

"His practical control seems to depend on the leeway he is given by an 'executive committee' which will, I believe, consist of himself, three Anglo American men – ETS "Ted" Brown, A Royden Harrison and Philip Harari – and one Diamond Royalties man, Dr PG Neethling, and which will take the practical everyday decisions on MDC's operations.

"It looks as if the swashbuckling days of absolute Collins control are over, and that the main driving force will now be the 'watchdog' executive committee, which will keep the board well posted on the day-to-day running of MDC."

In November 1965 Sammy arrived in Durban from Cape Town to supervise, as chief of Collins Submarine Pipelines Africa (Pty) Ltd, the pulling of a mile-and-a-half effluent pipeline at Umkomaas on the Natal South Coast. This was to be the first step in a R1 million scheme to clean up the seawater and beaches at Umkomaas which had been polluted by effluent from a cellulose factory.

Collins had established an office in Durban from which to administer his operations in Natal – including the contract referred to earlier to lay an offshore sewage outfall pipeline in the Bluff area for Durban Municipality – and this office was now run by Bill Webb, who had opted to remain with Collins after the De Beers takeover of Marine Diamonds.

The Durban City Council's decision to take Collins Submarine Pipelines off the sewage outfall contract after the damage caused to the pipeline by coastal storms, and the Council's subsequent withholding of payments to the company, had led to litigation.

Ultimately the company was awarded what Webb said was, at the time, the biggest insurance payout made to a South African company – R1,5 million.

Cape newspaper editor Piet Beukes lost a son in the ill-fated municipal outfall project. A professional diver, he was working for Collins's company when he got into difficulties in a deep dive, and drowned.

Sam Collins continued to make newspaper headlines, either about reported new projects he was looking into or in connection with litigation of one kind and another – but no longer concerning the mining of sea-diamonds.

He was cited either as the plaintiff or the defendant in a number of court cases at the Cape and in Durban.

In February 1968, Collins appeared in the Cape Town Magistrate's Court in connection with Capetex Engineering Works, his ship-repair business adjacent to Table Bay Harbour. He appeared as a director of Capetex, on 44 counts of failing to pay monies – R16 488 in all – due in terms of the Industrial Conciliation Act.

It was alleged that he had failed to make deductions from employees' salaries which should have been paid into funds controlled by the Industrial Council for the Iron, Steel, Engineering and Metallurgical Society of SA.

Six months later Collins sold this well-equipped but highly illiquid ship repair business to Murray and Stewart (Marine), at a price that was not revealed but which was said to have been big enough to wipe out Capetex's considerable debts. The firm had been the subject of numerous actions brought by creditors.

Also in 1968 a former mayor of Durban, Cyril Alexander Milne, was tried in the Supreme Court in Durban on charges of corruption and fraud. On the corruption charge he was, as a city councillor, alleged to have received or agreed to receive a fee from Collins to favour the interests of Collins Submarine Pipelines in connection with the contract for Durban's sea outfall sewerage installations.

Details of the alleged bribe referred to 35 000 shares made available to the ex-mayor in 1963, in Sea Diamond Corporation Ltd, when the market value of the shares was not less than 160 cents a share.

Milne denied there had been any bribery involved and that he had been expected to pay in full for the shares issued to him by Collins. He admitted, however, that he had not yet paid a promissory note for R17 500 – the value of the shares – in favour of Collins.

He also admitted that he had sold the shares in 1964 for R73 553, without having paid for them. He produced evidence to show that his net horse-racing losses in 1964/65 totalled R48 000.

Milne was found guilty on one count of corruption and three counts of fraud, and was sentenced to a total of 33 months in jail.

On the corruption charge, the judge said Milne had acted in "a most reprehensible and dishonourable manner" towards Collins. "It appears to me that even if he did not pay Collins because he was losing money at the races, he could have told Collins and asked him for an extension of time.

"Is it not more probable than otherwise that he did not do anything because he did not have to pay?"

Bill Webb says the shares were never meant as a gift from Collins; that the ex-mayor had applied for them and had been issued with the shares – but that he had failed to honour a written undertaking to pay for them.

Mystery venture on the Wild Coast

Sammy sparked media speculation when, in mid-1967, he turned up on Transkei's Wild Coast on a mystery venture. He was accompanied by a Briton, a German and two South Africans, and soon after their arrival at Lusikisiki it was rumoured that members of this team were either searching for diamonds, oil or the wreck of the *Grosvenor*.

None of the men would disclose the real purpose of the visit, then or later.

It was reported at the time that the team travelled daily to Port St Johns, where the ex-Marine Diamonds auxiliary vessel *Oom Kappie* and a barge had been operating a few miles off the coast. Divers had also been seen on the barge.

In 1968 Collins, on his return to Cape Town from a visit to his interests in the Persian Gulf, London and Texas, instituted an action for substantial damages following an article that had appeared in an Afrikaans Sunday newspaper suggesting that he had disappeared from South Africa with a young woman from the Cape, to escape his debts and problems.

Collins said in an urgent application for an order preventing the further printing, publishing and distribution of the newspaper that there was no truth in any of the suggestions made by it and that he at all times intended returning to South Africa, where he had considerable interests.

An out-of-court settlement was reached and Collins dropped his claim for damages after the newspaper concerned published a front-page apology and correction. This conceded that at the time the offending article was published, Collins was already back in South Africa.

The newspaper's publishers agreed to pay part of Collins's legal costs.

Servicing oil industry needs, in the Gulf

While retaining his home in Cape Town, Collins moved the focus of his interests and activities back to the Persian Gulf, where as chairman of Establishment Collins International, in Kuwait, he built up an organisation to service the needs of the oil industry in the Gulf.

To add to his small fleet of barges, tugs and other craft, he repurchased, in 1971, three of the vessels he and Consolidated Diamond Mines of SWA had used for diamond recovery off the SWA coast – the tugs *Collinsea* and *Chameis* (which had replaced the ill-fated *Collinstar*) and the former recovery barge *Pomona*, built by CDM.

The barge, needed by Collins mainly for the crew accommodation it offered, was towed to the Gulf by the *Collinsea*.

On one of his periodic visits to the Gulf Sammy became involved in an argument with a member of the crew of one of his service vessels, and it was reported that the rebellious crewman struck the Texan on the head with a spanner as he emerged from a hold in the vessel. He was knocked unconscious and had to be hospitalised.

In the Supreme Court, Cape Town, in December 1971, an order was granted authorising Shell (SA) (Pty) Ltd to effect service of process on Sam Collins, then of London, for the payment of R133 563.

Shell was given authority to effect service on Establishment Collins International, Collins Submarine Pipelines or a Mr AG Davies of 17b Curzon Street, London.

In an affidavit before the court it was stated that it had become impossible to effect service of a summons in Kuwait, and that all attempts to serve summons on Collins personally in London had proved unsuccessful. It had become apparent that the contents of the summons could be brought to his attention only at the Curzon Street address.

The papers before the court did not disclose the nature of the summons. A letter from the author of this book to Shell SA management, dated November 9, 1995, inquiring if the company's records showed whether or not this debt was ever paid by Collins, or on his behalf, brought the response more than two months later:

"Unfortunately we do not have any archived information on the matter."

From 1970, Sammy and Gigi had again made their home in London, but made frequent visits to South Africa to visit their many friends here. Former close associates still living at the Cape state that to their knowledge Collins made a point of clearing all outstanding personal and other debts.

As one of them put it: "He had a lot of trouble to face here in South Africa and in other parts of the world, but he was not the type of chap to duck his responsibilities. He was never afraid to face the music.

"Beneath his rough exterior was a very kind heart. Sammy Collins had courage and a very good brain. He was terribly loyal to South Africa, and in the ten years I worked with him I never knew him to tell an untruth in business. Rough as he may have been, I never knew him to do anything devious or dishonest. He was in reality a very straight person.

"I know for a fact that, because so many people took advantage of his generosity and helpfulness, Sammy was owed more than he owed. But he did not sue those who were indebted to him; to the last, he trusted them to pay him back."

In the 1971/72 edition of *Who's Who in the World*, Collins was shown as holding or having held 19 company chairmanships or directorships in various parts of the globe. Yet, at the peak of his career, he was said to have controlled more than 30 companies – including five in oil – and nearly 100 seagoing vessels.

"I'm not innarested in anythin' I don't control," he once said.

In mid-1969 Sammy, then aged 56, had hit the headlines again when it was reported from London that he was recovering from shock after he and a friend had fought off three armed and masked men who had broken into the Collins's Mayfair apartment.

Sammy was said to have used "a battery of judo chops and karate punches" to send the intruders running, and was referred to as "a former diamond-dredging magnate who once taught judo and karate to United States Air Force men".

Collins, Gigi and three friends had been drinking cocktails in the apartment in Berkeley Square when a man telephoned to say he would be delivering an urgent registered package.

When Sammy opened the door in response to a knock the three men, with stockings over their faces, burst in and, said Collins later: "I threw one of them against the door with all the force I could use. During the fight my friend broke his arm, but despite this he beat the hell out of one of the raiders.

"There was a terrific melee, and a judo battle that lasted several minutes."

People living in adjacent flats heard the scuffle and phoned the police, who arrived just after the raiders had escaped, empty-handed.

A report from London said: "The flat still had signs of the fierce fight yesterday. In the lounge, lined with Tretchikoff paintings, were stocking masks and rubber gloves used by the raiders..."

As he grew older, Sammy's years of hard-driving, hard living and heavy drinking began to take their toll, and while he was still operating his oil-rig service company from Dubai on the Persian Gulf in the 1970s his run-down condition became such that he had himself admitted to a sanatorium in Switzerland, for specialised treatment.

"Two-gun Collins – the man the world forgot..."

Then, in November 1978, came the news from London that Sammy Collins had died there, at the age of 65.

His wife Georgette was reported to have noticed, when she took him his early-morning coffee, that he was lying in an unaccustomed position. A doctor was summoned, and it was found that Sammy had died of a stroke – probably several hours earlier.

A London-sourced newspaper obituary that appeared at the time was headed: "Two-gun Sam Collins – the man the world forgot."

"He always carried two guns," the report said. "He wore a diamond-studded belt, cufflinks and tie-pin. And he shook the business world by dredging up diamonds from the bottom of the sea. But when he died last Monday it seemed the world had forgotten him.

"He was Sam Collins, the flamboyant Texan who in one day made nearly R4 million through the sale of shares in his company on the London Stock Exchange.

"On Friday he was cremated at Putney Vale in London, with hardly a mention in the Press. For, in direct contrast to the rip-roaring days when he was mining diamonds off Plum Pudding Island on the

South West African coast, and initiating the South African offshore oil search, he spent the last 10 years of his life almost in seclusion.

"Coming from a modest background and with only a high school education, Mr Collins made his name when he pioneered the construction of submarine pipelines – building his first when offshore oil was found in the Gulf of Mexico.

"The diamond business in Oranjemund lured him to South Africa – again in connection with the pipeline business. But it was there he conceived the idea of mining the seabed – with the world scoffing at his every move.

"But in 1961 he produced his first diamond from the sea – and his company, Sea Diamonds, took off. He also dabbled in oil exploration, but never really took up the options.

"He is survived by his wife Georgette, a son and two daughters."

At the time of Sammy's death, his son Sammy Junior was a regional manager of the Pelican Wharf restaurant operations in Texas. One of his daughters, Mrs Barbara Comstock, was living at Port Lavaca and the other daughter, Sister Stephanie, was attached to an order of Catholic nuns at Los Angeles. He also left five grandchildren and three great-grandchildren.

So ended the saga of Samuel Vernon Collins, adventurer and entrepreneur extraordinary; the boy from Beaumont, Texas, who came to Southern Africa and in four action-packed years added a new and exciting chapter to the story of gemstone mining in the sub-continent.

Where Barney Barnato had become part of the legend of the diamond rush at Kimberley near the turn of the century, Sammy Collins, sixty years later, ensured that his own name would forever be linked to the large-scale mining of diamonds from the sea.

The rusted wreck of *Barge 77*, at Panther Head on the desert shoreline of Chameis Bay, can perhaps be seen as a fitting memorial to an indefatigable fighter who braved the salty elements to translate into reality a belief he had shared with others, "in a land where men still dream".

CHAPTER FIFTEEN

Years later, "the Collins lamp is relit"

"I reckon they'll be mining diamonds off this coastline long after I'm not around anymore. There are enough diamonds down there to last a hell of a long time..."

Sam Collins, 1965

THIRTY years after Sam Collins faded from the scene, sea diamond mining off the West Coast of Southern Africa has become a thriving industry – with prospects of becoming ever-bigger, more sophisticated and more venturesome.

The De Beers group, which acquired Sam Collins's Marine Diamond operations in the mid-1960s, cut back on sea mining as such soon after the take-over and started instead an extensive exploration programme that lasted many years.

And when the group started its next phase, the commissioning of purpose-built mining vessels in the late 1980s, the interest of investors and other mining groups was roused again.

Now, using sophisticated equipment such as large-diameter tunnelling drills suspended from oilrig-type drill ships, and also remote controlled, bulldozer-like "crawlers" to recover diamonds from the seabed at depths of up to 200 metres, the De Beers group, drawing on its massive resources of money and men, has become an undisputed leader in this field.

The group has the greater part of the offshore mining concessions at its disposal, yet there is one other public company, operating on a smaller scale and using more conventional Collins-type methods and equipment which is also mining sea diamonds successfully – and profitably – off the West Coast. It does so as a friendly neighbour of De Beers.

The Ocean Diamond Mining group (ODM) has been involved in deep-water offshore diamond exploration and mining since its inception in late 1983, and has successfully developed its own expertise and technology in all three major aspects of the work – basic exploration, sampling and mining.

At the time of writing the group is operating two mining vessels and is, with the exception of De Beers, the leader in operating such craft.

The Ocean Diamond Mining group's recovery vessel Namibian Gem, *one of the latest ships working off Namibia.*

Several key members of ODM's operating staff once worked for Sam Collins, and cut their sea-mining teeth on the Marine Diamond Corporation's barges, in the 1960s.

ODM came into being when its founder chairman, Ivan Prinsep, turned over to it the sole and exclusive contracting right from Eiland Diamante Beperk – a subsidiary of the Trans Hex Group – which he had received personally from Trans Hex.

ODM thereby obtained the rights both to explore for and to mine diamonds from the territorial waters of 12 guano islands then owned by SA, off the coast of what was then South West Africa.

Sam Collins had conceived and created the Marine Diamond Corporation through the personal qualities for which he was known and admired – keen entrepreneurial skills, courage and dogged determination – and these same qualities in Ivan Prinsep brought ODM into existence years later.

While Prinsep came of a very different background from Collins – whom he first heard about from a taxi driver in Cape Town during an early visit to South Africa – he shared the Texan's tenacity and vision. And while the path followed by Prinsep in his bid for undersea mining rights was again very different from that taken by Collins, it was an equally fascinating one.

While Collins came from the world of oil and submarine pipelines, Prinsep's background is in international banking and investment. Born in India in 1927 of a British father and a Russian mother, he was brought up in England, went to the Royal Naval College at Dartmouth and served in motor torpedo-boats and a battleship.

He went on to become a Cambridge MA in history, after which he spent three years on Wall Street, training to be a stockbroker. After seven years in Canada where he worked with a famous financier, WA

Arbuckle, he spent six years in Trinidad as an investment banker.

Prinsep re-entered the financial world in Geneva in 1969, and there he married a Frenchwoman, Helene de Monbrison. They have lived in Switzerland ever since, with Prinsep commuting between his Swiss and Cape Town offices three or four times a year.

In the 1970s he became particularly interested in putting together international syndicates involving natural resources. He was one of a small group of people involved in a syndicate to exploit what became a huge natural gas field in Qatar in the Persian Gulf, and this gave him a taste for mineral resources.

A meeting with the famous Jacques Piccard

In the early '70s Prinsep was introduced to the renowned oceanographer Dr Jacques Piccard who, in 1960, with Lt Donald Walsh of the US Navy, had made the deepest-ever descent into the ocean – down the 11 000-metre Challenger Deep of the Marianas Trench in the Pacific. The descent was made in the US Navy's bathyscaphe *Trieste*.

"No air could be used as a lifting medium on that deep dive," says Prinsep. "Piccard explained to me the astonishing fact that, around 8 000 metres of depth, air will be so compressed that it will sink like the proverbial brick. Thus, the cigar-shaped vessel he used had contained oil, instead of air."

Since his days as a naval officer in experimental motor torpedo-boats, Prinsep has always had a fascination for technology, and he and Jacques Piccard became good friends. They discussed the advantages of searching for minerals rather than oil on the seabed – to save power, preferably close inshore.

Piccard suggested using submersible machines with huge independently-mounted wheels, which could be driven into the sea from beaches and then made either to "swim" on the surface or sink to the bottom for seabed mining.

He introduced Prinsep to an American geologist named Christensen who produced a map of the world indicating various mineral deposits. He suggested exploration in Liberia, but Prinsep opted for a trip to South Africa, which he first visited in 1977, with Christensen.

Initially the idea was to look for various minerals in this country, particularly titanium, but Prinsep's interest was then drawn to undersea diamonds.

He met a senior man in the South African Ministry of Mineral and Energy Affairs, Colin Border, and the information and encouragement he received from Border strengthened his resolve to try to get into undersea diamond mining.

Christensen introduced him to Hugo Richter of Cape Town – later to become Prinsep's partner – and to Dr Peter le Riche.

Dr Jacques Piccard, who in 1960 made man's deepest-ever descent into the ocean. In the '70s he helped Ivan Prinsep at the start of his sea-diamond mining venture.

"After I had told Richter what we were doing," says Prinsep, "we found ourselves lunching in the basement restaurant of the then Heerengracht Hotel, Cape Town, with Francois Hoffman who was chairman of Trans Hex, Terra Marina and other mining companies. Hoffie, as he is called, had made an excellent career in the mining world, in which he was greatly liked as well as being highly regarded.

"I told him that Jacques Piccard and I wanted to mine underwater, and asked if he could see his way to giving me a non-committal letter which would leave the door open for me for a year and allow me to attempt to stitch this thing together.

"After reflection, Hoffie appeared interested in the approach although he clearly considered it as highly risky. He said: 'If it's that non-committal, you can write the letter yourself, and I'll sign it'. When I drafted the letter in his presence the next day, he said I should also include the 12 South African guano islands with their territorial waters, on which a sister company to Terra Marina, Eiland Diamante Beperk, had held a lease.

"This happened at a stage when I was so green in this business that I didn't know the difference between a prospecting licence and a mining licence!"

Through his encouragement, Hoffie was to change the history of sea diamond mining, although at the time neither he nor Prinsep realised that.

Prinsep went at once to New York, to talk to International Nickel who had previously told him they had great interest in any prospect of working with Dr Piccard.

"They told me in New York and at their head office in Toronto that they considered the underwater project a highly interesting technological challenge, more than a geological gamble, as they already believed in the existence of substantial quantities of diamonds under those seas."

However, Prinsep adds, despite coming close to an agreement – with plenty of zeros on the numbers – he was told that the South African government's apartheid policy entirely inhibited any question of an investment in South Africa.

This same stumbling-block was to be encountered by Prinsep for many long and weary years, and prevented him from raising any serious amounts of capital to get the venture properly under way.

"South Africa is going to blow up", they said

"At that stage, in 1977, I had three arguments thrown at me. The first was that South Africa was going to blow up and that 'either we get our money back in twelve months or we're not interested'. The second was that while Jacques Piccard had certain ideas for undersea mining, these were still in his head, and not yet proven. And the third argument thrown at me – and the hardest one to answer – was 'if your idea was any good, De Beers would have done it long ago!'

"Ten years later Agip Minerale, the mining arm of the international giant Agip Oil, requested that they be permitted to send two experts to examine our work. The experts, a geologist and a mining engineer, reported back in Italy in very favourable terms.

"This time there had been even more zeros in the intended deal. But, with history repeating itself, the sad news was that apartheid prohibited an investment. Needing financial support, as we did, this came as a bitter outcome to the enthusiasm that had been shown by a major such as Agip."

Back in South Africa, in 1978, Prinsep and Richter who were anxious to keep their options open but with little time to do so, met a fishing-boat owner, Fanie van der Westhuizen, at Port Nolloth and they induced him to install a pump and a diver in his boat, the *Wakkerstroom*, and to "start pumping".

Van der Westhuizen agreed, good-humouredly, and it was also agreed that he and the partners in this venture go 50/50 on any diamonds found by the vessel.

"Fanie then asked where he must go to start pumping," says Prinsep. "This was at a time when everyone was working blindly and people used to laughingly say: 'Just put down your pipe close to that third seagull on the water, just next to those other two...'

"It was suggested his vessel go off the chimney-pipe near the De Beers beach mining plant at Kleinzee, so that was where we went, with De Beers and ourselves gazing at each other through binoculars.

"And then bingo! – we hit diamonds, and this became our very first commercial production."

Prinsep says this was a truly fantastic moment for everyone involved. "It was like winning the world cup, or being awarded a gold medal. All the sweeter in that some time before this, one of the large Canadian companies I had solicited had called Sam Collins by telephone and had heard him say: 'I don't know this man Prinsep, but if he says there are sea diamonds south of the Orange River Mouth, then forget it! There are none there, as I have found by sampling.'"

(Ironically, Prinsep's successful venture south of the Orange later earned him the compliment, from an internationally-recognised expert, Dr John Gurney, that he had "relit the Collins lamp" by renewing the search for sea diamonds).

In the first month in its new role, the *Wakkerstroom* produced 389 stones totalling 337,18 carats, with a value of R70 786 – and in those days the rand was worth well over one US dollar.

The Dawn Diamonds saga

Encouraged by this early and lucky strike, Prinsep and Richter progressed. They had used a shell company called Dawn Diamonds that Richter had registered but which until now had nothing in it. By 1980 Dawn, which was not itself a quoted company, was employing about 300 people – 75 of them divers, many of them former SA Navy men – and operating about 30 small boats.

The company profit for the financial year ending August 1980 was just over two million US dollars net, on a paid-up capital of less than R5 000.

While both Richter – who also had a company called Theron Holdings – and Prinsep had wanted a majority shareholding in Dawn, Prinsep allowed his partner to acquire a 51% interest in the newly-formed company on condition that the Prinsep group could reform their shareholding by taking partnership shares instead, "which was technically very important to certain corporate objectives."

Dawn Diamonds had three successive managing directors – Jack Walsh, Daniel Derwent and Hernus Kriel, who later became a cabinet minister and then, after SA's first democratic elections in 1994, Premier of the Western Cape.

In 1980 Richter decided, as a result of the seven-figure royalties that Dawn were paying it, to make a takeover bid for Trans Hex. This did not succeed, and indeed brought in other bidders – the winner being the giant Rembrandt group.

Unhappily, problems and great discomfiture followed within Dawn, and this culminated in a court action brought by their minority shareholders against Theron Holdings – the majority shareholders. An urgent application was heard by the late Mr Justice Berman of the Cape Supreme Court at his home at 11.30 at night – with the judge in his dressing-gown.

Ably assisting in this action against Theron Holdings were, notably, Steve Phelps and Anton Buirski. Phelps, a brilliant young accountant then working with Messrs Welby Stewart, untangled serious accounting matters while Anton Buirski, who then headed the law firm Buirski Herbstein and Ipp, was charged with the legal aspects. Harry Snitcher QC, a prominent member of the Cape Bar, was briefed for the plaintiffs.

Theron Holdings was finally put into liquidation in 1983.

Meanwhile, several entrepreneurs were wanting to pick up the pieces of Dawn Diamonds, which was finally acquired by Better Sales Limited, a small firm owned by companies controlled by financier and company director Geoff Grylls and others including Steve Phelps and, later, also by a company controlled by chain-store magnate Christo Wiese.

They eventually sold Dawn, after some tidying-up work had been done and after the company had produced an appreciable number of diamonds from what was known as Area 5A.

Following the action against Theron Holdings and its subsequent liquidation, which had been watched with interest by a large number of people in SA, Francois Hoffman, as chairman of the Trans Hex Group, once again entrusted to Prinsep, in his own name, an exclusive contract to exploit the territorial waters of the 12 islands held by a mining lease issued to Eiland Diamante by SA, in the late 1960s.

Eiland Diamante would receive a sliding royalty and a profit share above a certain level of profits, together with certain other rights, while Prinsep would have to provide all the financing needed including the spending of an agreed minimum each year.

Prinsep says the renewed confidence shown to him by Hoffman and the Trans Hex Group was "something I will treasure all my life".

He went to London with this contract and, after a few hesitant starts, was introduced by a private investment manager, Peter Hart, to Ian Cameron, senior partner of Panmure Gordon, an old-established firm of stockbrokers. He in turn arranged a meeting with the Hon John Baring, then chairman of Barings Bank.

Baring, assisted by Miles Rivett-Carnac who had had a brilliant career in the Royal Navy and then in banking – he later became deputy chairman of Barings – said "let's give it a go!" With the combined help of Barings and Panmure (Ian Cameron and Mark Henderson) Prinsep was able to raise 3,4 million US dollars, which was considered enough to launch a company and start work.

In this way Ocean Diamond Mining – so-named by Prinsep – came into being in late 1983. For various reasons, including tax, the company was incorporated in Guernsey, in the Channel Islands.

CHAPTER SIXTEEN

The proof of the Plum Pudding...

JUST before Ocean Diamond Mining came into being, Ivan Prinsep had bought an old coaster lying in Hout Bay called *Calypso*, which he turned over at cost to ODM.

The vessel was now fitted out for mining under the guidance of an expert named An Cornellisen, and sent to Plum Pudding Island.

"*Calypso*," says Prinsep, "was meant to be a clone of the *Ontginner*, which had been used in the late 1960s by Eiland Diamante with the near-legendary Hans Abel as dredgemaster. He had, in one month in an area he called the 'Glory Hole', recovered 10 000 carats at Plum Pudding, which was a tremendous feat."

Prinsep recounts that "old and dusty but factual" records obtained from the piled-up archives of Eiland Diamante of the accomplishments of Hans Abel and the *Ontginner* acted like "a map of where the treasure lies". His own eyes widened on reading them, he adds, "and the reaction of the early investors was similar.

'Without those figures of previous recoveries I doubt whether our backers would have had the courage to go into the project. Yet here was actual proof – proof of the Plum Pudding, if you like – that diamonds had been recovered."

Barings Bank, when approached for backing, had baulked at the idea of starting the venture with submersibles not yet built by Dr Piccard, says Prinsep. "They considered such an unproven concept – however intellectually appealing – to be too venturesome for anyone they would wish to contact.

"And that was the reason we decided to go back to the systems that Sam Collins had used in the '60s. We are still in most respects using those systems – albeit much improved – effectively today."

Geoff Grylls, with experience gained through his acquisition and then practical management of Dawn Diamonds, was invited to become ODM's first managing director, and he remained in this position until 1991. Steve Phelps, who had fought so stoutly and effectively to protect the interest of the minority shareholders in Dawn, was invited to join the board, as was Tony Buirski who had successfully led the legal fight.

"Geoff put in a lot of stalwart, constructive work, ably assisted by Jimmy Volkwyn," says Prinsep.

"I recall Geoff commenting one day: 'This really is a complex and difficult business; there's no point of reference, no textbook you can open to see how to mine sea-diamonds. Managing this entirely new ocean-going venture is rather like trying to read a newspaper in a heavy wind!'"

Grylls, a former Springbok swimmer, recalls that Captain Ronnie Grice "skilfully drove *Calypso* and its dredging operations by the seat of his pants".

Grice, with An Cornellisen and Jimmy Volkwyn, a young chartered accountant Grylls had brought into the ODM operation, achieved extraordinary results in their efforts to motivate men having to work in a hostile environment with limited technology at their disposal, he adds.

In 1983, at the time that ODM came into existence, diamonds were selling at 98 US dollars a carat, as against the $37 (or R27) being paid in Collins's time. The benchmark price used at the time of writing (1996) is about 200 dollars a carat.

In August 1985, on the eighth voyage of the *Calypso*, under Captain Grice, ODM was joined by Tom Kilgour who, as mentioned earlier, had in the 1960s been in charge of technical aspects of Sam Collins's Marine Diamond Corporation. He had also worked for the De Beers group, at Oranjemund.

By the time it was decided to sell *Calypso* the vessel had, up to November 1986, collected diamonds totalling 6 051 carats. She never quite made profits although coming very close on occasion. The vessel had proved to have too small a capacity for production and was too large and wrongly set up for sampling only.

"At this stage," says Prinsep "one of our directors, the late Dr Henry Meyer – representing the interests of a shareholder called Harry Winston & Co – suggested that to review our work to that date, Dr Donald Sutherland of Placer Analysis Ltd, Edinburgh, should be engaged to make studies and give recommendations. This was duly done.

"While the company was regrouping, so to speak, it was of vital importance to obtain a cash flow – as in any business, of course. To that end I suggested that we should try to exploit the shallow waters around Possession Island, using the same methods which Dawn had used in previous years – divers.

"So the company obtained permission to set up a small base camp on the island, taking great care not to disturb the bird-life. It then acquired a small wooden fishing vessel, the *Simonsberg*, which served it well by producing 3 688 carats between January 1987 and March 1989 and bringing home that much sought-after cash flow. The key divers who made it happen were Mike Bouch, Anthony Buchanan and Charlie Bosse.

"Another calculated risk had, with the merciful aid of Providence, paid off!"

In 1989, ODM registered a company called ODM Namibia in Windhoek, in anticipation of South West Africa gaining independence the following year. This was a difficult decision at the time, as Possession Island where ODM was busy mining was still South African territory, and diamond gravels could not be moved from one national territory to another.

A treasure-trove in the "Lobster Claw"

The diamonds recovered at Possession came almost entirely from the north-east corner of the sea off the island, in a spot bounded by a multitude of dangerous rocks which Prinsep nicknamed "The Lobster Claw".

He recalls that Mike Bouch, who had been a diver in the North Sea, had to venture deeper and deeper off the island, and reported that the diamondiferous gravel was still continuing at a depth of 31 metres.

"But, regrettably, we decided we had to stop mining any deeper as we did not have the decompression facilities essential for work at those depths."

In December 1989, ODM bought a slightly larger steel-hulled vessel called *African Bounty*, but this was not a success and was eventually sold. Before this, in February 1987, Gershon Ben-Tovim, an Israeli with plentiful experience in engineering and marine work, had come into the picture.

He approached Prinsep, saying he had just ended a contract and badly wanted a new contract for his ship, *Johanna II*, or he would most probably have to close down his operations. ODM eventually agreed that Ben-Tovim work a specified area, on a 50/50 basis on diamonds produced.

Meanwhile Prinsep was doing the rounds of mining houses in South Africa, looking for much larger sums of capital. While failing in his approach to the major mining houses he found one smaller company that showed a real interest. This was Benguela Concessions Limited (Benco), which had been put together by Dr John Gurney, a University of Cape Town scientist with a world-wide reputation as an authority on kimberlites and related subjects.

Gurney had acted as a consultant for Prinsep in 1983/4 in estimating diamond content in the waters of Plum Pudding Island and all the other 11 islands farther to the north. He had estimated a diamond content of at least 2 million carats in the area covered by the mining lease of Eiland Diamante.

Since then, Gurney had formed his own company.

In 1987 ODM had been delighted to receive a letter from the Minister of Mineral and Energy Affairs congratulating the company on the success of its application two years previously for two new areas within South African waters. These were designated 6C and 14C and, says Prinsep, Dr Gurney was "particularly bullish on 6C, which bordered Benco's own 6B – both areas of which he believed to contain a vast quantity of diamonds".

An exciting and important deep-water test in area 6C by ODM – down to 137 metres and about 15 km offshore – had produced some diamonds which, while not many in quantity, definitely proved certain geological postulations, to the delight of ODM.

Also, until then, no-one other than De Beers itself had recovered diamonds by managing to work either at such a distance from shore, or at such a depth.

Ivan Prinsep proudly wears, as a tiepin, one of the very first stones (1,98 carats) recovered on that first deep-water trip. He recounts that a South African subsidiary of one of ODM's larger shareholders – Ellermans – managed by UOA Brown of Cape Town, "courageously provided the welcome financial support of over R2 million for this notable achievement".

Both Norman Gillman and Tom Kilgour of ODM were deeply involved in this venture, using the vessel *Deep Salvage* (Captain Peter Wilmot) from Cape Town Harbour.

Benco agreed to give ODM R3,5 million in cash and to spend R20 million over five years on exploring those two C areas. Benco thereby obtained an interest of 49,9% in these areas, which could later be increased to 66,7% under certain conditions.

With this important agreement in hand, Dr Gurney evoked the interest of the Australian-based BHP (Broken Hill Proprietary), said to be the world's biggest mining conglomerate and, after certain conditions had been agreed to, Gurney succeeded in getting BHP to come in as a partner with Benco.

Never ceasing in his attempts to raise capital, Ivan Prinsep all of a sudden found himself being introduced by one of his directors, Steve Phelps, to a major insurance company, Sanlam. There, he met a senior portfolio manager, Prieur du Plessis, and also Andre C Louw, Sanlam's senior manager, mining investment research. Louw had previously held a senior position in BP Minerals.

Sanlam agrees to invest in ODM

Detailed negotiations were then entered into with Sanlam, ODM agreeing to comply with certain conditions, the main points being that ODM should convert its debt into ordinary shares and that the whole of ODM be brought to South Africa from Guernsey and then go public, with a listing on the Johannesburg Stock Exchange.

"Sanlam," says Prinsep, "to its everlasting credit agreed to accept my proposal which was for a non-cumulative 6 percent convertible preference share instead of the much more usual high coupons attaching to either debentures or preferred shares.

"After being initially taken aback by this unusual proposal, Sanlam on reflection agreed that it made sense not to burden, at the outset, a small company like ODM with high interest charges while it was still exploring and growing, but instead to give it every chance to mature and provide capital gains to its backers.

"The ODM shareholders – which of course include Sanlam itself as holders of the convertible preference shares – have every reason to be thankful for the wisdom of Sanlam's management in accepting that reasoning."

Andre Louw later called on Prinsep and said he would like to get back into active mining, and in August 1991 he took over as managing director of ODM, in succession to Geoff Grylls who had served ODM long and well since 1983.

Before Sanlam came into the picture, the ODM operation had been scaled down considerably and was in a virtual "hold" situation with Grylls helping to keep it going while at the same time running other companies in which he held key positions.

"We had been unable to go back to the London markets following PW Botha's infamous 'Rubicon' speech," he recalls. "This had effectively torpedoed our efforts to raise the capital we needed, and for four years we had to go into this 'hold' pattern, with my own function having been scaled down by myself, to that of being more of a caretaker than an MD."

(President PW Botha, with the eyes of the world on him, made a major speech in Durban City Hall in July 1985, in which he was widely expected to announce sweeping reforms in his government's apartheid policy and a positive and radical change of political direction for SA. When these expectations were not met, Botha was said to have "failed to cross the Rubicon" – a failure that led to a slump in the local currency and to a large-scale disinvestment by foreign companies).

Grylls continues: "Over all those years I had done the job of helping Ivan and ODM much more out of love and the sheer appeal and interest of the thing rather than money. In fact ODM, which took a small office on the floor where a number of other companies with which I was involved were located, was helped to the point of almost being subsidised by us.

"It was only when Sanlam came into it that the company could continue on an entirely different footing, and at that stage it had become apparent that, with the way technology was moving, the financial specialists had to make way at the top for geologists, engineers and others with the necessary technical and practical know-how.

"We could all have given up hope at that stage..."

"I had told Ivan Prinsep that there was not much more that I could do, other than help him keep the company's head above water. We had the mining concessions but we lacked the financial backing to exploit them, so we had to take things a day at a time until we had scaled down to almost nothing.

"We could all have given up hope at that stage, and it was only Ivan's drive and vision that kept us going although none of us knew what the end result was likely to be.

"Ivan, in inevitable moments of near despair – because 'money' was not listening – always believed that the industry would mature; that the mining world would eventually acknowledge not only that the sea was going to be the next big source of gemstone diamonds but also a profitable source, and that the big players would come into the market and play a major role financially.

"Today, he has proved to be absolutely right."

In 1991, at the time that Andre Louw chose to get back into mining and was appointed MD of Ocean Diamond Mining, Sanlam told him they considered he might be making a mistake with this move, as he was leaving an extremely important post.

However, says Ivan Prinsep, "events have once again proved another correct decision. Andre Louw has accomplished a hard task admirably."

With Sanlam coming into the ODM picture, the mining company made another deal with Ben-Tovim, because it clearly needed to have production, not just assets. Ben-Tovim came back into ODM's life with a vessel he had recently acquired called the *Lucky D*, which worked in a joint venture for ODM from September 1991 to March 1993 and which made a side-trip to prospect in Alexander Bay in 1991. He had, says Prinsep, "contrived to finance the acquisition of his ship from one of ODM's own shareholders!"

The joint venture was not a happy one, and was ended in April 1993. The vessel, which had been fitted with ODM's mining equipment, was taken over one-hundred percent by ODM in May 1994, and renamed *Oceandia*.

In the same year, with the aid of its stockbrokers, Simpson McKie of Johannesburg, and its international consultants, James Capel and Co Inc of London, ODM raised some R60 million through a public offering.

With this money the company was able to purchase from Tidewater Oil in Britain the 2 500-ton *Regal Service*, a former North Sea oil-rig supply vessel. Its conversion into a mining unit by ODM's own people, in Cape Town, cost about R33 million overall. This was accomplished by Andre Louw and his hard-driving team in record time, and under budget.

Registered under the South African flag, and renamed *Namibian Gem*, the new mining vessel started production in April 1995.

The result was a huge increase in turnover and profitability. A year later, the company was able to report record profits showing that the *Namibian Gem* had lived up to all its promises. Earnings for the financial year ending in March 1996 were up to R6 million, from R106 000 the year before. Moreover, the company still had no debts and a cash hoard, ready for expansion, of nearly R30 million.

The smaller *Oceandia* had also performed well, acting mostly as a sampling vessel providing areas for the larger *Namibian Gem*. The ODM team under the able leadership and drive of its MD, Andre Louw, was, in Ivan Prinsep's words, "achieving wonders".

ODM policy is to employ as many Namibians as possible in its operations, and it has made a R100 000 grant to the Rössing Foundation to help a school in Luderitz at which seagoing personnel can be trained. Early in 1996 the Foundation handed the school over to the Namibian government.

After agreement had been reached between Pretoria and Windhoek that the formerly South African-owned port of Walvis Bay and the neighbouring guano islands would be incorporated into Namibia, companies with any vested interests in these areas were assured they would not be worse off than before.

ODM's offshore mining rights extended 15-fold

But then followed over two years of great anxiety. Growing awareness of what was happening, stemming from publicity by De Beers, ODM and others caused a large number of companies to enter the lists to try and obtain properties in Namibian waters. (It is rumoured that there were over 50 applicants in all).

Indeed, more than one competitor openly scoffed at the idea that ODM – despite being the first recognised applicant for the extension of the waters off the 12 islands – would succeed in its application. Some of those in the mining world were becoming aware of the enormous potential of these areas, where the prehistorically-formed and now-drowned beaches could contain untold amounts of gemstones.

ODM was much impressed when Barry Davies, BHP's manager in Cape Town, reported that BHP had decided not to apply for the island extension areas as his company knew that ODM had been the first applicant and that BHP had no desire to interfere with or endanger ODM's application in any way.

"One does not forget such things," says Prinsep.

In mid-1996, with the tide turning strongly in ODM's favour, David Gleason, Editor-at-Large of the *Financial Mail*, wrote under the headline "Shades of the A-team", that: "Investors who kept the faith are reaping handsome rewards from Ocean Diamond Mining, the independent West Coast marine diamond exploration and mining company.

"After years of struggle, the long-term strategic plan has turned the company around and set it on a path to regular profits. MD Andre Louw says: 'This is the end result of a plan which is coming together. ODM is now demonstrating consistent, high-quality production.'"

Then came a significant milestone in the history of ODM. On June 12, 1996, the group's stockbrokers disclosed through the media that they had been authorised to announce that the Namibian government had granted extensive new sea diamond prospecting rights off the Namibian coast to ODM's wholly-owned subsidiary, Island Diamonds Limited (IDL) – a company that ODM had purchased from Trans Hex and others.

The brokers also pointed out that IDL already held Mining Licence 36 which covered the "territorial waters" of 12 islands in the Luderitz area up to the three-nautical-mile limit off the Namibian coast.

The new award, covering a total of 5 700 square kilometres, would increase about fifteenfold the area of the ODM group's Namibian prospecting and mining rights.

The brokers disclosed that ODM had postulated the existence of at least two ancient drainage systems inputting large quantities of gem-quality diamonds to the sea in the region granted to it.

"In taking this decision," says Prinsep, "Namibia's government, and particularly its Ministry of Mines and all those associated with it, most honourably followed the rules Namibia had itself set as to treating applications of this nature.

"But surely playing a part in its decision was an awareness by the government that ODM had worked hard, efficiently and with professional competence in developing this important new industry within their country, and was employing increasing numbers of Namibians as well as paying growing royalties to the nation."

The *Mining Journal* in London, reporting on the new award under the headline "ODM Lands Jackpot", said in its June 14, 1996 issue:

"A clearly delighted ODM chairman, Ivan Prinsep, told *Mining Journal* this was of 'prime geological importance' and represented a massive boost to the company's prospects."

The journal added that: "Plans are already advanced to acquire another vessel now that, after years of waiting, the Namibian government has made its move."

In South Africa, the new award prompted David Gleason to write a follow-up article in the *Financial Mail*. Headed "Moving into the Big

Time" this said the award had put "a different complexion" on the ODM group's prospects.

Gleason quoted the group's financial director in Cape Town, Jo Consul, as saying the new area was "in the middle of one of the world's richest diamond sectors; it will make a very big difference to our potential".

Gleason wrote that until mid-1996, ODM had intended introducing an additional mining vessel, to come into service only late in the 1998 financial year. "Consul now expects that timetable to be advanced substantially.

"Nor does he discount the possibility that three or even four more vessels may be pressed into action".

Shortly after the new award to ODM had been announced, it was reported that diamond sales by De Beers' Central Selling Organisation for the first half of 1996 had jumped 8.2% on the same period in 1995, to $2,7 billion, outstripping expectations and setting the group on the road to record full-year sales.

The sales, underpinned by rising retail demand, particularly in the jewellery market, raised the prospect of the CSO lifting the quota on suppliers for the first time in two years.

All in all, the prospects for Southern Africa's diamond industry – marine diamonds particularly – had never looked better. The picture had changed dramatically from that of two decades earlier.

Ivan Prinsep, who deplores the "lack of tax-driven incentives (compared with Canada, Sweden and many other countries) for South African resident investors", recalls that when he first came to SA, in 1977, virtually nothing was happening in the sphere of sea diamond mining.

"The spirit had gone out of the venture. Eiland Diamante had just taken an awful blow, and the Terra Marina group were doing nothing. De Beers was working hard at sea on endless surveying and sampling, but was definitely keeping very quiet indeed about its findings, and its views of the ocean's prospects.

"This thick blanket of silence had its intended effect on the investment world, which assumed that because of lack of information or news from the largest company in the diamond industry, the ocean had little or no prospects for serious mining. Only about three years ago did De Beers start drawing back the veils on this subject.

"On an early visit to this country I was being driven somewhere in a taxi – in Cape Town – and the driver asked what I was doing here. When I told him he replied: 'Ah, then you'd better study the story of Sam Collins!'.

"The taxi-driver then started telling me all about Collins. He obviously knew a lot about this man who, so I soon found out, had been something of a folk hero here.

"And years later, Dr John Gurney startled me with the compliment that I had 'relit the Collins lamp', by restarting the drive for sea-diamonds with the evident success of Dawn Diamonds.

"And when one looks back today at the venturesome and admirable Sammy Collins, one can see that he had the bad luck to come into sea-diamond mining too early. To the surprise of most, he pulled an awful lot of diamonds from the sea – estimated at 1,5 million carats.

"But, at around R27 a carat, and even with the rand then worth over 1,35 US dollars, it simply was not a paying proposition at the time.

"One is reminded, in *David Copperfield*, of Mr Micawber's dictum:
'*Annual income twenty pounds, annual expenditure nineteen pounds, nineteen and sixpence, result happiness; annual income twenty pounds, annual expenditure twenty pounds, ought and sixpence, result misery*'"

Andre Louw, managing director of an expanding Ocean Diamond Mining Group.

CHAPTER SEVENTEEN

Sammy, it seems, only scratched the surface...

WHEN Sam Collins and others prospected for diamonds in the sea three decades ago they were considered crazy; yet today it is clear that those bold pioneers did no more than scratch the surface of a vast treasure chest in the ocean.

It is now generally considered that the world's largest and most valuable resource of gem diamonds is contained not only in the currently exposed – and once-drowned – marine gravel beaches along the Namibian and South African west coasts but also in submerged beaches adjacent to the present coastline.

These stones are quite remarkable in that, over the whole of the huge area being mined, they are largely consistent in size, colour, cutability and high quality.

With vastly improved and developing technology, marine diamond mining has become a highly sophisticated industry – way beyond the dreams of those earlier entrepreneurs.

Geologists have estimated that as many as 10 billion carats were originally liberated by erosion over a period of 100 million years from numerous diamondiferous kimberlites in the interior of Southern Africa.

It is believed that of this, something like 3 billion carats – of which 95 percent are gemstones – have over the ages been transported by ice and river into what is now the sea area. The reason for this extraordinarily high proportion of gemstones, say the experts, is that only the best-quality stones survived the glacial and fluvial transportation process to the coast.

As Ivan Prinsep of Ocean Diamond Mining puts it: "If one accepts the scientists' estimates, these make all the gold reserves in South Africa pale by comparison!

"If one assumes 3 billion carats to be an estimate that will one day be proved true, the 'diamond-to-gold' comparison can be calculated as follows:

"Present price of gold, ca $400 an ounce; present price of sea diamonds, ca $200 a carat. Therefore, two carats being worth one ounce of gold, 3 billion carats (or 1,5 billion ounces of gold equivalent) equates to some 43 000 tons of gold.

"South Africa mined less than 600 tons of gold last year (1995)."

There may be an impression among laymen, says Prinsep, that diamonds can be found lying on the floor of the sea "like potatoes or carrots, in a field. But this is absolutely not so.

"Since the diamonds were delivered millions of years ago to what is now the sea, the oceans have transgressed and regressed several times, for a swing of something like 400 to 500 metres up and down, in relation to the present sea-level. These large changes in sea-levels were of course caused by global warming and cooling which, although much in the news today, is nothing novel to geologists.

"In this process beaches, rivers, deltas and lagoons were formed and every time the sea stopped for a while – say for half a million years – it made areas of very high turbulence along coastlines, with constant wave action.

"This coastal turbulence has had the effect of sifting marine matter and putting it into its own layer of specific gravity, with like matter being brought together by these powerful natural forces.

"There's a standing joke in the club bar of a golf course I used to play on at Fishers Island off the Connecticut coast, that if people drove their balls into the sea there and did not recover them, they'd be able to come back in six months and find all the balls in the same spot!

"So what we have found since the Collins era is that these 'standstill' areas as they are called – the coastal areas where the sea stood still for very long periods – are the valuable targets. These are the selected areas we aim for today."

Before, at and since Collins's time it has been noted that the marine diamonds found off the West Coast are of exceptional colour, clarity and quality – a quality in constant demand at the commercial end of the jewellery market.

"What has to be understood about West Coast diamonds," says Prinsep, "is that although they are not big – they average half a carat or less – they are extremely fine stones.

"If you take a half-carat diamond in the rough and finish up with a stone of 0.25, or 0.27 in the cut state, this so happens to be the size of the small, modest, but extremely high-quality diamond in an engagement ring – or anniversary ring – which is in great demand around the world. And this is a very comfortable position for a diamond mining company to be in."

Prinsep laughingly quotes a diamond dealer in Antwerp, with long experience of handling gemstones, who had said:

"As long as we have romance of the heart in the world there will be a demand for diamonds. A man first buys a diamond to give to his fiancée. Then he buys one to give to his mistress, and then he buys a third diamond to give to his wife when she finds out about the mistress – and this business will be going on forever!"

The 'Jules Verne' plan of a former U-boat captain

Since Collins's time, a marine mining capability has been developed by De Beers and others which is way beyond anything comparable elsewhere in the world.

A De Beers marine mining executive has described methods and equipment being used today as "real Jules Verne technology", which could eventually be extended to mine other high-value minerals such as gold, silver and platinum from the sea floor.

He was thinking, as Dr Jacques Piccard had suggested to Ivan Prinsep in 1977, that the best way was to work from the sea-bottom itself.

In 1995 a latter-day Jules Verne came on the marine mining scene with a plan to use former military submarines to dive for diamonds at great depths off Southern Africa's West Coast.

Herbert Werner, one of the last surviving World War II German U-boat captains was, at the age of 76, planning to hunt for diamonds at depths of 200 to 600 metres – depths that have so far been commercially inaccessible – in a multi-million rand project.

This American-domiciled real-estate millionaire, whose book on the U-boats, *Iron Coffins*, became a bestseller translated into 14 languages, spoke of revamping surplus British Oberon and Russian Kilo-class patrol submarines with an array of electronics and robotics to turn them into mining control rooms.

Submarines and remote-controlled underwater drilling rigs that could operate at any depth would, he said, deliver rich harvests of oil, gas, gold, cobalt and manganese nodules from the ocean floor to a resource-hungry world by early in the 21st century.

Herbert Werner's Florida-based company, Werner Deep Offshore Research, was granted diamond prospecting rights in a 4 300 sq km concession to 120 km west of the Orange River delta in August 1995, by South Africa's Department of Mineral and Energy Affairs. The search area, south of the extension of the South African-Namibian border, abuts the successful De Beers concession which produced more than 400 000 gem-quality carats in 1994.

Although De Beers does not use divers, the waters in which it is working being much too deep, there is still some diver-operated work going on off the West Coast. While Ocean Diamond Mining does not itself use divers, it has diving sub-contractors working for it around the Namibian islands.

The larger companies are using their suction pipes in a robotic manner to obtain diamondiferous gravel, both from the floor of the seabed and from its crevices and potholes.

Namco, a Vancouver-based company, at the time of writing is still in the survey and sampling stage, while Benco has not yet started meaningful mining. But ODM finds itself in the fortunate position of being able to mine at a profit while continuing to explore its large concessions.

Ivan Prinsep, having initially wished to mine from the floor bottom with the ideas of Dr Jacques Piccard, has some sympathy for the ideas of Herbert Werner, in his sub-sea plans.

It could well be that the giant company De Beers Marine, Herbert Werner, ODM, Benco and Namco, and all others of a like mind will, in one way or another, succeed in writing a whole new chapter in the story of underwater mining – as the doughty Sammy Collins did in the 1960s.

They are all pioneers, and who was it who said:

"The sands of the desert are strewn with the bones of pioneers, but without those brave men, where would we be today?"

Tom Kilgour, formerly Sam Collins's technical chief and now operations manager of Ocean Diamond Mining.

ADDENDUM

WITH reference to the final three chapters of this book, featuring the post-Collins era – the early days with Dawn Diamonds and then the birth and growth of the Ocean Diamond Mining Group – ODM's founder and chairman, Ivan RM Prinsep, writes:

"Acknowledgements are due to many people who have not been specifically named in the foregoing pages.

"While none have been forgotten, no book could be long enough to contain all their names and to recount what each did. In many and varied ways, they have each and all helped build something from dream to reality.

"Not in any order of importance – because they were all important – are: the late Dr Fred Collender (for his technical and intellectual support); Pierre Pictet, Swiss banker (for seed capital when needed most); all our loyal and patient shareholders; Jo W Consul, our devoted and far-seeing financial director: W Ian L Forrest, CA (who served for many years as deputy chairman of ODM, proving to be an invaluable friend of the company; the late Lord (Nick) Hardinge of the Royal Bank of Canada; Drayton, the diamond hunter and unsung hero who at Hondeklip Bay early in the whole saga improved the impeller of the 6-inch diamond pumps to the great benefit of all; the Bateman group under Roger Falls; Edward Sniders MBE in Geneva (for much backing and encouragement over many years); Stuart Lyle, Andrew Phillips, Michael C Abbott and Michael J Foster; Raymond Mallach (who in Dawn Diamonds's early days was a partner with Buirski Herbstein & Ipp before starting his own practice; the then advocate Selwyn Selikowitz (now a judge); our auditors, Don Bowden of Deloittes, and lawyers (Tony Hardy of Fairbridges); our director Neil Hoogenhout, representing Eiland Diamante, and then our first Namibian director, Peter Koep, and a whole host of faithful fellow workers both ashore and afloat, geologists, shipmates and technicians in both Namibia and South Africa – plus one or two specialists from overseas.

"Finally, a word of appreciation of the fair play evidenced by the prevalent rule of law of both the governments of the republics of South Africa and of Namibia. This has provided a foundation for confidence and investment. Without such a foundation, on which investors have been able to build, nothing can or will happen.

"All this must never be merely taken for granted, but recognised and taken note of. So, all the many government officials met in Springbok, Cape Town, Johannesburg, Pretoria and Windhoek, and Luderitz, who knew us and know us and who may happen to read this book, must please accept these words of thanks.

INDEX

A

A Grue of Ice 22
A Twist of Sand 22
Abbott, Michael 171
Abel, Hans 87-92, 157
African Bounty 159
African Salvage Corporation 31
Agip Minerale 154
Agip Oil 154
Albertyn, Piet 102
Anglo American Corporation .. 20, 27-9, ...37-8, 54, 67, 138
Anglovaal 37, 41, 58-9, 138
APB 45 .... 96, 98
Arbuckle, WA 152
Argus Oil Exploration Co 74, 143
Arpione 31, 33
Atherstone, Dr Guybon 19
Atkinson, Kevin 82
Atlantiese Diamant Korporasie 45,139
Auckland 36
Aunty Lil 105

B

Barge 77 ... 50-57, 65, 69, 71, 82, 87-8, 93, ...97-100, 109, 117, 149
Barge 111 .... 82, 93, 95, 119, 140
Baring, the Hon John 156
Barings Bank 156-7
Barnato, Barney 11, 20, 149
Barrow, Brian 63
Bateman Group 171
Ben-Tovim, Gershon 159, 162
Benguela Concessions (Benco) ...... 159, ...169-70
Berman, Mr Justice 155
Better Sales Ltd 156
Beukes, Piet 14, 16, 53-7, 144
Bloomberg, Abe .... 18, 30-1, 37, 41, 76, ...130, 133
Bombay Port Trust 15
Bonuskor 132
Border, Colin 152
Bosse, Charlie 158
Botha, D 92
Botha, PW 161
Bouch, Mike 158-9
Bowden, Don 171
Boyes, Lorenzo 19

BP Minerals 160
Broken Hill Pty (BHP) 160, 163
Brouwer, "Fido" 64
Brown, ETS (Ted) 27, 139, 144
Brown, UOA 160
Buchanan, Anthony 158
Buirski Herbstein & Ipp 156, 171
Buirski, Anton 155-7

C

Caltex 135
Calypso 157
Cambridge University 151
Cameron, Ian 156
Cape Aero Club 103
Cape Times 57, 63, 97, 105, 130
Cape Town Art Centre 82
Capetex Engineering Works 144
Carr, Jock 101, 111-3
Carstens, Capt Jack 24
Catacombs, The 84
Central Selling Organisation .... 49, 59, ...138, 165
Chameis 105, 146
Chameis Bay Holdings 139
Chan 105
Christensen 152
Collender, Dr Fred 171
Collins Construction Co 15
Collins, Frank & Ellie 13
Collins, Georgette (Gigi) .... 31, 56-7, 64, ...81, 97, 126, 129, 147
Collins, Sammy Jnr 64, 91, 149
Collins Submarine Pipelines . 15, 30, 60, ...71, 75, 111, 144-5
Collinsea 107, 146
Collinstar 106, 108
Colpontoon 96, 105-10, 119, 140
Comstock, Mrs Barbara 149
Congella 44
Consolidated Diamond Mines (CDM) ...
..................24, 27, 30, 38, 48, 58, 99, 146
Consul, Jo 165, 171
Cooke, Ainsley 102
Cornellisen, An 157
Cortes 105, 108
Cottonpicker 83, 101-2
Cranko, Captain R 92
Cullinan Diamond 121
Cypress 105

D

Dannhauser, Valerie 40
Daryl's 84
Davies, AG 147
Davies, Barry 163
Dawn Diamonds 155-7, 166
De Beer, Johannes N 19
De Beers Group 18, 20, 23, 27, 30, 37,
.....49, 58-60, 105, 120, 127, 135, 138,150,
....160, 163, 169
De Beers Marine (Debmar) 115
De Beers Mining Company 19, 21
Deep Salvage 160
Deloittes 171
De Monbrison, Helene 152
De Olim, Johnny 38
Department of Geography, Wits University 27
Derwent, Daniel 155
Deutsche Diamanten Gesellschaft 23
Deutsche Koloniale Gesellschaft 21
Diamantkus 96, 99-111, 116,
....119, 140-2
Diamond Fields of SA 21
Diamond Mining & Utility Co 135
Diamond News & SA Jeweller 70, 99
Diamond Producers' Assn 20
Diamond Regie 24
Diamond Royalties & Holdings 41,
....138, 144
Diamond Syndicate 24
Die Banier 82
Die Burger 81
Die Huisgenoot 116
Die Landstem 57
Diederichs, Dr Nico 56
Doms, J 41
Draper & Plewman 19
Drayton 171
Du Preez, AP 29, 37, 41, 45, 76,
....132-3, 135
Dugan, James 17
Duineveld Beleggings 56, 132-3,
....135
Duyzers, Hans 115
Dunkelsbuhler & Co 20
Du Plessis, Prieur 160
Durban City Council 60-1, 144

E

Eiland Diamante 134, 151, 153, 157,
....159, 165, 171
Ellermans 160
Emerson K 8, 42-54, 60, 92, 100, 108,
....113-5, 125

Establishment Collins International ....
....146-7
Evans, Bill 28, 60, 74

F

Fairbridges 171
Falls, Roger 171
Federale Mynbou 132
Federale Volksbeleggings 132
Financial Mail 57, 100, 141, 163
Financial Times 94
Forrest, W Ian L 171
Foster, Michael J 171
Foulis, Capt Don 118, 128
Foulis, Capt George 8, 44-6, 60, 79,
....97, 105, 128
Fridell, Bob 102

G

General Mining & Finance Corporation
....................18, 37, 41-2, 57-9, 93-4, 138
Getty, John Paul 132, 135
Gianni, Laury 102
Gibbs, George 102
Gillman, Norman 160
Glasscock, Billy 60
Gleason, David 163-4
Globe Engineering Works 93, 98
Gobel, Captain Karl 106, 108-9
Godfrey, Graham 107
Government Metallurgical Laboratories
....93
Grapow, Capt "Okkie" 107
Great Eastern 17
Green, Lawrence 25-6, 34
Greenshields, Bob 87, 121, 124
Grice, Capt Ronnie 158
Grosvenor 145
Grylls, Geoff 156-7, 160
Gurney, Dr John 155, 159-60, 166
Guthrie, George 74

H

Haak, Jan 54-5
Hackenschmidt, George 31
Harari, Philip 144
Hardinge, Lord (Nick) 171
Hardy, Tony 171
Harker, Captain Graham 107, 110
Harris, Captain CJ 136
Harrison, A Royden 144
Harry Winston & Co 158
Hart HN 41
Hart, Peter 156

Harvey, Wynn 60, 74
Haselau, Gary 77-9, 86, 125, 129
Heard, Anthony 57
Henderson, Mark 156
Hendry, Jock 111
Hoffman, Francois (Hoffie) 12, 27,
.... 42, 45-9, 75, 87, 125, 135, 142,
.... 153, 156
Hoogenhout, Neil 171
Horne, Andrew 84
Horne, Joe 84, 87, 90, 95
Hossfly 101, 103
Human, Johnny 102

I

Illingsworth, Bob 102
Industrial Council, Iron & Steel 145
Industrial Diamonds Group 135
International Nickel 153
Ipp, Danny 41, 139-40
Ipp, J 41
Iron Coffins 169
Island Diamonds Ltd (IDL) 164

J

Jacobs, Erasmus S 19
James Capel & Co Inc 162
Jars, Fred 87, 91
Jenkins, Geoffrey 22
Johanna II 159
Johannesburg Stock Exchange 160
Johnson, Brian 102
Jonker Diamond 27
Jules Verne 169

K

Kagan, Sidney 67-70, 140
Kahan, ME 132, 135
Kailey, Emerson 14, 36-7, 41-2,
.... 94, 127
Kaptein, Aat 81
Karibib 10, 38
Keeble, Peter 16-18, 30-1, 37, 41-3,
.... 78, 130
Kennedy, Jackie 129
Kennedy, President JF 129
Kilgour, Tom 67, 93, 117,
.... 158, 160
Kimberley Central Company 20
Klipbok 137
Koep, Peter 171
Koloniale Bergbau Gesellschaft 23
Kriel, Hernus 155

L

League of Nations 43
Le Riche, Dr Peter 102, 152
Levinson, Olga 11
Louis (Collins's chauffeur) 64, 66
Louw, Andre C 160-4
Louw, Dr MS 132
Lucky D 162
Lyle, Stuart 171

M

Macmillan, Harold 127
Macmillan, Lady Dorothy 127
Malherbe, Fasie 54
Mallach, Raymond 171
Manuel, E 92
Manzi Fé, Baron 32
Marauder, HMS 8, 42
Maree, BD 132
Marina 104, 118-9, 123
Marine Diamond Corporation 37,
.... 41-4, 49, 58-9, 67, 71-2, 74-5, 87, 93,
.... 101, 127, 133, 138, 150
Marine Products 36
Martens, FL 41
Martin, Catherine 83
Martin, John 83
Martin, Marilyn 81-5
Martin, Norman 83
Mashona Coast 105
McLaughlin 105
Maxime's 64, 66, 84
Mazy Zed 22
Menell, SG (Slip) 37, 41
Merensky, Dr Hans 11, 25
Metropolitan Golf Club 82
Meyer, Dr Henry 158
Meyer, PH 132
Middle Wits 138
Middle, Carl 102
Milne, CA 145
Milner, Bunny 140
Mining Journal 59, 164
Mitchell, JW (Bill) 41, 44, 64, 101
Mohawk Airlines 102
Moresby-White, J 130
Murray & Stewart (Marine) 145

N

Namibian Gem 151, 163
Namibian Minerals Corporation
(Namco) 169-70
National Institute of Metallurgy 93
National Party 130

Nautilus 35
Navigator's Den 84
Neethling, Dr Piet 29, 41, 45, 76,
.... 132-4, 139, 144
Nienaber, Nicki 89

O

Oakes, Golly 102
Ocean Diamond Mining Group 150,
.... 156-9, 162-4, 167
Ocean Science & Engineering 136
Oceandia 162-3
Oceanographic Research Committee ....
.... 125
ODM Namibia 159
Ogilvie Thompson, Julian 139
Ontdekker 1 133
Ontginner 157
Oom Kappie 105, 146
Oppenheimer, Harry 37, 51, 54, 58-9,
.... 138-9
Oppenheimer, Sir Ernest 18, 20, 23
Orama Holdings Ltd 139
Ordeal by Water 17

P

Panmure Gordon 156
Panther Head Investments 139
Parliamentary Angling Club 129
Pharoah, Capt Hugh 74, 102-3
Phelps, Steve 155-7, 160
Phillips, Andrew 171
Piccard, Dr Jacques 152-4, 169-70
Pictet, Pierre 171
Placer Analysis Ltd 158
Pleitz, Albert 43
Pomona 140, 146
Pretorius, Louw 116
Prinsep, Ivan 7, 151, 159, 163-4,
.... 167-8, 170-1

R

Range, Dr 21
Reader's Digest 17
Redelinghuys, Col Rex 18, 86, 112,
.... 122
Redgrave, Eddie 101, 103, 128
Regal Service 162
Rembrandt Group 27, 155
Rhodes, Cecil John 19-20
Richter, Hugo 152
Rijger 46
River of Diamonds 22
Rivett-Carnac, Miles 156

Rockeater 136-7, 141
Rodin, Leroy 74
Rössing Foundation 163
Rothschild Family 20
Royal Bank of Canada 171
Royal Naval College, Dartmouth 151
Rudd, Charles 19
Ruffneck 101

S

SA Handbook 20
SA National Gallery 83
SA Navy 155
SA Shipping News 98
Sable Room 64, 84
Safmarine 44, 107
Sandow, Eugen 31
Sanlam 134, 160-1
Schipa 44-5, 57, 95, 101, 103, 117-8,
.... 123, 128
Schlesinger, John 81, 127
Scholtz, Daantjie 55
Scott, Col Jack 18, 37, 41
Sea Diamond Corporation 41, 48,
.... 58, 60, 94, 101, 133, 138-9
Seagulls Restaurant 82
Selikowitz, Judge Selwyn 171
Seymour 105
Shell in Industry 52-3
Shell SA (Pty) Ltd 146
Sibley 43, 90, 92
Simonsberg 158
Simpson McKie 162
Simpson, Prof ESW 125
Slabber, Conrad 91
Smith, Capt Alan 110
Smith, Mrs Pat 110
Smith, Neil J 118
Smith, Wilbur 23
Smithsonian Institution 103
Sniders, Edward 171
Snitcher, Harry 156
Societa Ricuperi Marittimi 32
Southern Diamond Corporation 52,
.... 58-9
Spes Bona Mining Co 132
Standard Bank of SA 37, 69
Stauch, August 11, 21
Steenkamp, Lukas 54
Stephanie, Sister 64, 149
Steyn, Marais 130
Strydom, Capt GHF 18, 41, 45, 54,
.... 105, 133
Suidwes-Afrika Prospekteerders . 40, 139
Sunday Express 117
Sutherland, Dr Donald 158

175

T

Tegniek 36
Telenews 79
Terblanche, Jurie 107
Terra Marina 87, 132-4, 153
The Diamond Hunters 23
The Great Iron Ship 17
Theron Holdings 155
Thesen's Steamship Co 128
Tidal Diamond Corporation 27
Tidewater Oil Co of America 135,
....162
Titan Products 31
To the River's End 25
Trans Hex Group 27, 135, 151, 153,
....155-6, 164
Tretchikoff, Vladimir 56, 86, 126,
....148
Trieste 152
Trust Bank 134
Tsumeb Corporation Ltd 99
Twyford 33

U

Unie Vleis 143
United Party 130
University of Cape Town 33, 125,
....159
University of Durban Westville 83
University of SA 83
University of Witwatersrand 83, 93
US Army Air Force 148
US Navy 152

V

Van der Westhuizen, Fanie 154
Van Diggelen, Tromp 31
Van Heerden, Bill 102
Van Staden, JW 130
Van Zyl, Gerrie 36, 39-41
Van Zyl, Senator Johannes 36, 39,
....40-1
Vartsos, Socrates 31, 38, 63, 65
Veedol Minerals (Pty) Ltd 135
Vema 126
Verwoerd, Dr HF 56, 130
Viljoen, Daan 96
Vivier, Johann 10, 29, 36-8, 41, 45-6,
....52, 67, 80
Vivier, Mrs Babs 39, 80
Vogue Academy 81
Volkwyn, Jimmy 157-8
Vooruitzicht 19
Vorster, John 130

W

Wagner, PA 21
Wakkerstroom 154-5
Walsh, Jack 155
Walsh, Lt Donald 152
Webb, Bill 12, 18, 26, 42-3, 46,
....89, 143-5
Webster, A 41
Weinreich, Audrey 81-2
Welby Stewart, Messrs 156
Wellington, Prof John 27
Werner Deep Offshore Research 169
Werner, Herbert 169-70
Weskus Mynbou 54-6, 132
West Wits 57
Where Men Still Dream 34
Wiese, Christo 156
Williams, Alpheus F 18, 23
Williams, Captain DD 42
Williamson Mines 98
Willow 105
Wilmot, Captain Peter 160
Wilson, Dave 35
Wisboom-Verstegen, JW 108
Wodehouse, Sir Philip 19
World Court 134
Worthwhile Journey 31
Wright, Dr JA 54

Y

Yolanda 46, 105

Z

Zwarte Zee 107